SpringerBriefs in Applied Sciences and Technology

Computational Intelligence

Series Editor

Janusz Kacprzyk, Systems Research Institute, Polish Academy of Sciences, Warsaw, Poland

SpringerBriefs in Computational Intelligence are a series of slim high-quality publications encompassing the entire spectrum of Computational Intelligence. Featuring compact volumes of 50 to 125 pages (approximately 20,000–45,000 words), Briefs are shorter than a conventional book but longer than a journal article. Thus Briefs serve as timely, concise tools for students, researchers, and professionals.

More information about this series at http://www.springer.com/series/10618

Seyedali Mirjalili · Jin Song Dong

Multi-Objective Optimization using Artificial Intelligence Techniques

 Springer

Seyedali Mirjalili
Torrens University Australia
Fortitude Valley
Brisbane, QLD, Australia

Jin Song Dong
Institute for Integrated
and Intelligent Systems
Griffith University
Brisbane, QLD, Australia

Department of Computer Science
School of Computing
National University of Singapore
Singapore, Singapore

ISSN 2191-530X   ISSN 2191-5318   (electronic)
SpringerBriefs in Applied Sciences and Technology
ISSN 2625-3704   ISSN 2625-3712   (electronic)
SpringerBriefs in Computational Intelligence
ISBN 978-3-030-24834-5   ISBN 978-3-030-24835-2   (eBook)
https://doi.org/10.1007/978-3-030-24835-2

This Springer imprint is published by the registered company Springer Nature Switzerland AG
The registered company address is: Gewerbestrasse 11, 6330 Cham, Switzerland

To my father and mother

Preface

This book focuses on the most well-regarded and recent nature-inspired algorithms capable of solving optimization problems with multiple objectives. First, the book provides preliminaries and essential definitions in multi-objective problems and different paradigms to solve them. It then provides an in-depth explanation of the theory, literature review, and applications of several widely used algorithms. The algorithms are Multi-objective Particle Swarm Optimizer (MOPSO), Multi-Objective Genetic Algorithm (NSGA-II), and Multi-objective Grey Wolf Optimizer (MOGWO).

Brisbane, Australia   Dr. Seyedali Mirjalili
July 2019   Prof. Jin Song Dong

Contents

Acronyms

EA	Evolutionary algorithm
GA	Genetic Algorithm
PSO	Particle Swarm Optimization
GWO	Grey Wolf Optimizer
SA	Simulated Annealing
MOPSO	Multi-Objective Particle Swarm Optimization
MOGWO	Multi-Objective Grey Wolf Optimizer
NSGA	Non-dominated Sorting Genetic Algorithm
PF	Pareto Optimal Front

Chapter 1
Introduction to Multi-objective Optimization

1.1 Introduction

In the field of Artificial Intelligence (AI), search algorithms have been popular since their invention. A search algorithm is typically designed to search and find a desired solution from a given set of all possible solutions to maximize/minimize one or multiple objectives. Depending on the mechanism of a search method, this set of solution can be searched entirely or partially. A search algorithm starts with an initial state (solution), and the ultimate goal is to find a target state (solution). Note that there might be multiple targets in case of multi-objective search that will be discussed in a later section. One of the main challenges in the field of AI is that the set that should be searched by a search algorithm exponentially grows proportional to the size of the problem and the number of objectives. This was not an issue in the past when the problems were less complex and challenging. These days, however, this issue should be addressed when solving a wide range of problems.

1.2 Uninformed and Heuristic AI Search Methods

One of the most well-regarded classifications in the area of AI search algorithms divides such methods into two classes: uninformed and informed. In the former class, an algorithm does not have any additional information about its distance to the goal state. This means that an uninformed search is only aware of the problem definition and should make decisions without knowing the quality of each solution during the search. Each action in such methods is equally good, and a lot of people refer to them as blind search methods. This results in being computationally more expensive and slow when solving large-scale problems. Some of the most popular blind search methods are Breadth-First Search (BFS), Depth-First Search (DFS), brute-force search, and Iterative DFS.

© The Author(s), under exclusive license to Springer Nature Switzerland AG 2020
S. Mirjalili and J. S. Dong, *Multi-Objective Optimization using Artificial Intelligence Techniques*, SpringerBriefs in Computational Intelligence,
https://doi.org/10.1007/978-3-030-24835-2_1

In the latter class, informed search, there is additional information that shows an estimate of the distance between the current state and the target state. Such methods are often called heuristics search algorithms [1] as well. A heuristic algorithm leverages a heuristic function to make educated decisions when taking actions, therefore, each action is not equally good as opposed to uninformed search methods. The heuristic function allows evaluating each action and chooses the most promising ones. As the result, informed search methods skip a large portion of the search set (space) and are less computationally expensive than uninformed search algorithms. They can be applied to large-scale problems as well, which is one of the main reason why they are of the most popular search methods lately. Some of the most well-regarded heuristic algorithms are greedy search [2], hill climbing [3], and A* [4].

An example of an uniformed and an informed search can be seen in Fig. 1.1. The problem is to find the highest peak on a terrain using a one-wheeler robot. The terrain is divided into 144 nodes as it can be seen as a grid in Fig. 1.1. The uninfomed method is an uninformed search and goes through all the states. The path highlighted in this figure shows that this search goes through each node row by row. Although the path stopped on the highest peak in this figure, brute force algorithm keeps searching until the last node and compared them all to find the highest one. This guarantees finding the best solution for any terrain, but it is computationally expensive.

Figure 1.1b shows an informed (heuristic) search method, in which all the possible neighboutin points from a given node is evaluated first and the best one (with the maximum elevation) is chosen. This algorithm is called hill climbing, which is one of the most conventional heuristic methods. If we follow the best solution each time, Fig. 1.1 shows that th robot will reach the highest hill with only nine steps. This algorithm is fast similarly to other heuristics. However, it is not complete and if we stat from a wrong initial solution, the navigator will lead the robot to hill that are not the highest hills on the terrain. This is shown in Fig. 1.2. This figures shows that changing the initial position of robot highly change its performance despite the higher speed of its navigation algorithm compared to the brute-force search algorithm.

1.3 Popularity of AI Heuristics and Metaheuristics

As the above example showed, a heurisitc algorithm finds "good" solutions in a reasonable time. This originated from the educated decisions that such methods make to chose more promising states from a given set of possible states. In the example, choosing one neighbor out of four each time using their altitudes cuts down the size of search space significantly. The key point here is that they are unreliable in finding the best possible state (solution) for a given problem. They are normally used when we cannot afford having informed search methods. A heuristic algorithm is problem specific. For instance, A* uses a heuristic function that calculates the spatial distance to the goal state so it can be used to traverse a tree. This mean that a problem should be represented using a state space tree [5] to be solved by the A* algorithms.

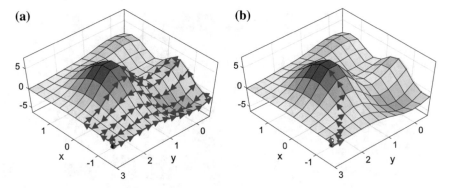

Fig. 1.1 a Uninformed search: a brute force search require to check all the possible states (locations) on the terrain, which is equal to 144 (14*14) states. The path that leads to the highest hill is highlighted. **b** Following the location with the highest altitude in the neighbored of each solution is a heurisitc algorithm and requires nine steps only to find the highest peak on the terrain

Fig. 1.2 The heuristic algorithm in Fig. 1.1 will lead the robot to a hill that is not the highest hill on the terrain

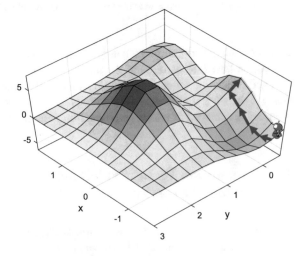

Problem dependency of heuristic algorithms motivated researchers to design metaheuristics [6], which make few assumption about the problem. This makes them more applicable than heuristics algorithms. Metaheuristics mostly employ stochastic operators to be able to efficiently explore the search space in order to find near-optimal solutions. As opposed to deterministic algorithms, such stochastic techniques find different solutions in each run. Another advantage of metaheuristics is the gradient-free mechanism. They do not need to calculate the derivative of the problem to find its best solution. In other words, metaheurstics consider a problem as a black box.

The above-mentioned advantages have led to an increasable popularity of metaheurstics in a wide rage of fields [7]. The majority of metaherustics follow the same model of search. They start with a set of initial solutions (states) for a given problem. This set may include one or multiple solutions. In the former class, the algorithm

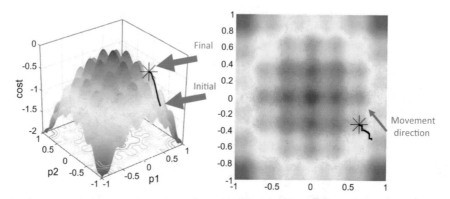

Fig. 1.3 A simple hill climbing algorithm, in which the algorithm starts from a initial point and chooses the best neighbouring solutions each time. This figure shows that such a behaviour, which is called exploitation, may result in finding a sub-optimal (often called locally optimal) solution. This originates from the fact that the algorithm always selects the best solution, whereas sometimes it needs to use a "bad" solution to avoid trapping in locally optimal solutions

is called a single-solution metaheuristic. In the latter case, the algorithm is called population-based. Regardless of the number of solutions in the set, it is constantly changed depending on the algorithm's structure. It means that algorithm iteratively changes the set and evaluates the solutions. This process is stopped when a certain accuracy is achieved or after a given maximum number of iterations.

1.4   Exploration Versus Exploitation in Heuristics and Metaheuristics

It is worth mentioning here that stochastic operators in metaheursitcs increase their exploration, which refers to the discovery of different regions in a search space. Exploitation is opposed to exploration where an algorithm tries to search locally instead of globally. To better understand these two conflicting behaviours in metaheuritics and other search methods, Figs. 1.3 and 1.5 are given. Figure 1.3 shows a simple hill climbing algorithm, in which the algorithm starts from an initial point and chooses the best neighbouring solutions each time. This figure shows that such a behaviour, which is called exploitation, may result in finding a sub-optimal (often called locally optimal) solution. This originates from the fact that the algorithm always "goes up" in Fig. 1.3. To get to the highest peak, however, the algorithm might also need to step down sometimes to pass the valleys and discover new and possibly better hills. Therefore, pure exploitation is good for linear problems (see Fig. 1.4).

As opposed to exploitation, an algorithm takes 'bad' actions occasionally to find more promising regions and avoid getting trapped in sub-optimal solutions of a

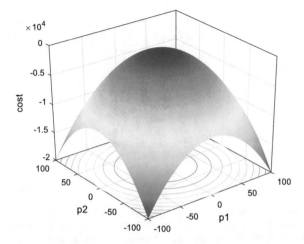

Fig. 1.4 An example of a problem with linear search space. To solve such problems, there is no need to perform exploration since exploitation leads us to the best solution. This is because there is no locally optimal solutions

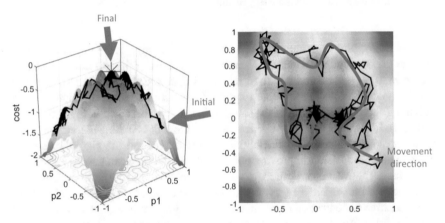

Fig. 1.5 Simulated Annealing (SA) is used which gives a certain probably to choosing downhill steps. This causes exploration of the search space as the arrow shows. Such a behavior allows the algorithm to avoid a large number of sub-optimal solutions

search space. This is shown in Fig. 1.5. In this experiment, Simulated Annealing (SA) [8] is used which gives a certain probability to choose downhill steps. This causes exploration of the search space as the arrow shows in Fig. 1.5. Such a behavior allows the algorithm to avoid a large number of sub-optimal solutions. To find an accurate solution, however, the algorithm needs to exploit the search space at some point. In SA, this is done by decreasing the probability of choosing downhill decrease proportional to the iteration. Overall, exploration is required when solving non-linear problems and should be accompanied with exploitation.

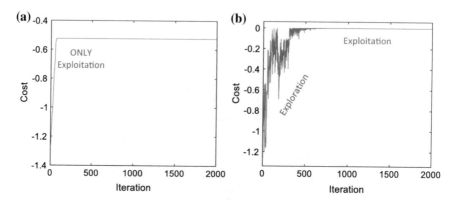

Fig. 1.6 a This is the convergence curve of the Hill Climbing algorithm, in which there is no fluctuations in the curve, and the algorithm quickly converges towards a non-improving points in the initial stages of the search process. **b** The Simulated Annealing algorithms shows a different convergence patterns. In the first 500 iterations, the solution in each iteration faces abrupt changes in the objective value which is due to taking downhill steps that leads to worse cost. However, this is changed after near 600 iterations, in which the algorithm starts the exploitation phase and finds the tallest peak around the most promising solution found in the last iteration

Exploration and exploitation can be observed by looking at the fluctuation in the altitude of the solution (which is often referred as cost or fitness) in the landscape represented in Figs. 1.3 and 1.5. To further investigate such behavious, Hill Climbing and Simulated Annealing are both required to search the global optimum in 2000 iterations and the fluctuation of the solution in each run is shown in Fig. 1.6.

Figure 1.6a shows that convergence curve of the Hill Climbing algorithm. In can be observed that there is no fluctuations in the curve, and the algorithm quickly converges towards a non-improving points in the initial stages of the search process. As discussed above, this is desired for a linear problem due to the lack of sub-optimal solutions. As illustrated in Fig. 1.6b, however, the Simulated Annealing algorithms shows a different convergence patterns. In the first 500 iterations, the solution in each iteration faces abrupt changes in the objective value which is due to taking downhill steps that leads worse cost. However, this is changed after nearly iteration 600, in which the algorithm start exploiting the exploitation phase and finds the tallest peak around the most promising solution found in the last iteration.

The above discussion and figures showed that exploration and exploitation are both required and should be balanced when searching for the global solution of problems with non-linear search spaces. Several figure showed how an algorithm might be trapped in a sub-optimal solution. To see the impact of performing only exploration, Fig. 1.7 is given. This figure shows that the coverage of the search space is very high when the exploration is at its highest level. Figure 1.7a and b show the algorithm searches more than 2/3 of the search space. Figure 1.7c illustrates the changes in the cost values, in which the fluctuation can be seen from the first to the last iterations. The issue here is that there is no systematic convergence and the final

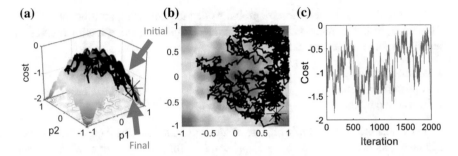

Fig. 1.7 **a** and **b** show the algorithm searches more than 2/3 of the search space. **c** illustrates the changes in the cost values, in which the fluctuation can be seen from the first to the last iterations. The issue here is that there is no systematic convergence and the final solution obtained (see the arrow in Fig. 1.7a) is even worse than the case where the algorithm only performs exploitation (see Fig. 1.3)

solution obtained (see the arrow in Fig. 1.7a) is even worse than the case where the algorithm only performs exploitation (see Fig. 1.3).

Another popular classification of metaheuristic is evolutionary, swarm-based, and physics-based algorithms [9]. In the first class, an algorithm mimics evolutionary phenomena in nature. For instance, Genetic Algorithms (GA) [10] mimics the way that organism's best genes propagate through generations to increase the chance of survival. In the second class, the problem solving of a swarm that originates from the social intelligence of individuals. For instance, Ant Colony Optimization (ACO) [11] mimics the process of finding the shortest path from a nest to a food source in an ant colony. In fact, ACO simulates the local interactions between ants using pheromone that allows solving such a complex problem in a very dynamic environment. In the last class, physics-based algorithm, physical phenomena are the main source of inspiration. A popular algorithm in this class is Simulated Annealing, which simulates the annealing process (heating and controlled cooling) of a material to increase its crystal's size.

1.5   Different Methods of Multi-objective Search (Optimization)

Regardless of the class metaheuristics, they are mostly considered as optimization algorithms. They have been largely employed in the literature to find the optimal values for the parameters of a problem to maximize or minimize an objective (often called as cost or fitness) function. If the problem has one objective, the algorithm is called a single-objective optimization. In a single-objective problem, there is one global solution to find. In a multi-objective problem, however, there are more than one objective, which might be in-conflict. This makes the process of optimization a lot more difficult than single-objective problems due to addressing multiple objectives. There are different methods in the literature of multi-objective meta-herustics to handle multiple objectives as follows:

- A priori methods
- A posteriori methods
- Interactive methods

In a priori method, multiple objectives are aggregated into a single objective. For instance, this process can be done using a weight for each objective, which allows us to define the importance of objectives. After the weighted aggregation, a single-objective optimization algorithm can be employed to find the global optimum. There are several drawbacks with this method that will be discussed in a later chapter.

In a posteriori method, the multi-objective formulation of the problem is first maintained. An algorithm (often multi-objective) is then employed to find the best trade-offs between the objectives. This leads to finding a set of solutions called Pareto optimal solution set. This method has its own pros and cons as well that will be discussed later on. However, it is worth mentioning here that a decision making process is required after the optimization process to choose one of the solutions.

In the interactive methods, decision makers are involved during the optimization process. This method of multi-objective optimization is often called human-in-the-loop optimization due to the direct impact of a human on the space searched by an algorithm.

1.6 Scope and Structure of the Book

In this book, the preliminaries, essential definitions, and state-of-the-art optimization algorithms to solve multi-objective optimization problems will be presented. It is tried to present several algorithms to show how multi-objective optimization is done using the three above-mentioned paradigms. The rest of the book is organized as follows:

Chapter 2 provides preliminaries and essential definitions of multi-objective optimization specially. A priori, a posteriori and interactive multi-objective optimization are also discussed in details. Two conventional metaheuristics including Multi-objective PSO (MOPSO) and multi-objective GA (NSGA-II) are presented and analyzed in Chaps. 3 and 4. Chapter 5 covers the recently-proposed multi-objective Grey Wolf Optimizer (MOGWO).

References

1. Kanal L, Kumar V (eds) (2012) Search in artificial intelligence. Springer Science & Business Media, New York
2. Resende MG, Ribeiro CC (2003) Greedy randomized adaptive search procedures. In: Handbook of metaheuristics. Springer, Boston, (pp 219–249)

3. Yuret D, De La Maza M (1993) Dynamic hill climbing: overcoming the limitations of optimization techniques. In: The second Turkish symposium on artificial intelligence and neural networks. Citeseer, (pp 208–212)
4. Liu X, Gong D (2011) A comparative study of A-star algorithms for search and rescue in perfect maze. In: 2011 International conference on electric information and control engineering. IEEE, (pp 24–27)
5. Esposito F, Malerba D, Semeraro G (1993) Decision tree pruning as a search in the state space. In: European conference on machine learning. Springer, Berlin (pp 165–184)
6. Boussaïd I, Lepagnot J, Siarry P (2013) A survey on optimization metaheuristics. Inf Sci 237:82–117
7. Osman IH, Kelly JP (1997) Meta-heuristics theory and applications. J Oper Res Soc 48(6):657–657
8. Kirkpatrick S, Gelatt CD, Vecchi MP (1983) Optimization by simulated annealing. Science 220(4598):671–680
9. Mirjalili S, Lewis A (2016) The whale optimization algorithm. Adv Eng Softw 95:51–67
10. Goldberg DE, Holland JH (1988) Genetic algorithms and machine learning. Mach Learn 3(2):95–99
11. Dorigo M, Blum C (2005) Ant colony optimization theory: a survey. Theor Comput Sci 344(2–3):243–278

Chapter 2
What is Really Multi-objective Optimization?

2.1 Introduction

Optimization refers to the process of finding an optimal set from the set of all possible solutions for a given problems. An algorithm is normally developed call optimization algorithm to find such a solution. Regardless of the specific structure, such algorithms required to compare two solutions at some stage to decide which one is better. An objective function (often called fitness, cost, etc) is used to evaluate the merit of each solution. In case of using one objective function, the proglem is single objective and the solutions can be compared using relational operators.

On the other hard, when we want to optimize more than one objective functions, the solution cannot be compared with relations operators. In this case, we might optimizer each objectives independently, but the problem is that the objectives are often in conflict. To solve such problems using optimization algorithms there are different methods in the literature [1]. This chapter covers different classes of multi-objective optimization techniques and essential definitions in this field.

2.2 Essential Definitions

Before introducing multi-objective optimization, it is important to know how a single-objective optimization problem is formulated. In the field of optimization, most of the definitions are presented for minimization problems without the loss of generality. This means that changing relational operators to their opposite leads to formulate a maximization problem. In this book, we provide both definitions to highlight the differences. A single-objective optimization can be formulated as a minimization problem as follows:

$$Minimize: \quad f(\overrightarrow{x}) \tag{2.1}$$

© The Author(s), under exclusive license to Springer Nature Switzerland AG 2020
S. Mirjalili and J. S. Dong, *Multi-Objective Optimization using Artificial Intelligence Techniques*, SpringerBriefs in Computational Intelligence,
https://doi.org/10.1007/978-3-030-24835-2_2

$$\text{Subject to}: \quad g_i(\vec{x}) \geq 0, i = 1, 2, ..., m \tag{2.2}$$

$$h_i(\vec{x}) = 0, i = 1, 2, ..., p \tag{2.3}$$

$$lb_i \leq x_i \leq ub_i, i = 1, 2, ..., n \tag{2.4}$$

where $\vec{x}$ is a vector that stores all the variables ($\vec{x} = \{x_1, x_2, x_3, ..., x_{n-1}, x_n\}$) for the problem, n is number of variables, m is the number of inequality constraints, p is the number of equality constraints, lb_i is the lower bound of the i-th variable, and ub_i is the upper bound of the i-th variable.

A single-objective optimisation can be formulated as a maximization problem as follows:

$$Maximize: \quad f(\vec{x}) \tag{2.5}$$

$$\text{Subject to}: \quad g_i(\vec{x}) \geq 0, i = 1, 2, ..., m \tag{2.6}$$

$$h_i(\vec{x}) = 0, i = 1, 2, ..., p \tag{2.7}$$

$$lb_i \leq x_i \leq ub_i, i = 1, 2, ..., n \tag{2.8}$$

where $\vec{x}$ is a vector that stores all the variables ($\vec{x} = \{x_1, x_2, x_3, ..., x_{n-1}, x_n\}$) for the problem, n is number of variables, m is the number of inequality constraints, p is the number of equality constraints, lb_i is the lower bound of the i-th variable, and ub_i is the upper bound of the i-th variable.

In the above formulations, there is a vector that store multiple variables (often called parameters or decision variable) store all the variables of the problem. This vector is passed into the objective function that returns a number as the results. In a multi-objective problem, however, there are more than one objective function to be called using the vector. We store those Multi-objective optimization. A multi-objective optimization can be formulated as a minimization problem as follows:

$$Minimize: \quad \overrightarrow{F(\vec{x})} = \{f_1(\vec{x}), f_2(\vec{x}), ..., f_o(\vec{x})\} \tag{2.9}$$

$$\text{Subject to}: \quad g_i(\vec{x}) \geq 0, i = 1, 2, ..., m \tag{2.10}$$

$$h_i(\vec{x}) = 0, i = 1, 2, ..., p \tag{2.11}$$

$$lb_i \leq x_i \leq ub_i, i = 1, 2, ..., n \tag{2.12}$$

where $\vec{x}$ is a vector that stores all the variables ($\vec{x} = \{x_1, x_2, x_3, ..., x_{n-1}, x_n\}$) for the problem, n is number of variables, m is the number of inequality constraints, p is the number of equality constraints, lb_i is the lower bound of the i-th variable, and ub_i is the upper bound of the i-th variable.

A multi-objective optimization can be formulated as a maximization problem as follows:

$$Maximize: \quad \overrightarrow{F(\overrightarrow{x})} = \{f_1(\overrightarrow{x}), f_2(\overrightarrow{x}), ..., f_o(\overrightarrow{x})\} \tag{2.13}$$

$$Subject\ to: \quad g_i(\overrightarrow{x}) \geq 0, i = 1, 2, ..., m \tag{2.14}$$

$$h_i(\overrightarrow{x}) = 0, i = 1, 2, ..., p \tag{2.15}$$

$$lb_i \leq x_i \leq ub_i, i = 1, 2, ..., n \tag{2.16}$$

where $\overrightarrow{x}$ is a vector that stores all the variables ($\overrightarrow{x} = \{x_1, x_2, x_3, ..., x_{n-1}, x_n\}$) for the problem, n is number of variables, m is the number of inequality constraints, p is the number of equality constraints, lb_i is the lower bound of the i-th variable, and ub_i is the upper bound of the i-th variable.

As discussed above, a new operator is required to compare two solutions when we are dealing with multiple objectives. This operator was proposed by Pareto called Pareto optimal dominance [2]. According to this operator, one solution is better than another if it shows equation objective values on all objectives and at least is better in one of the objectives. An example is price and quality. These two objectives are in conflict when a shopper is looking for an item. As the quality goes down the price goes down too. In reality, most shoppers are looking for the best trade-offs between these two objective. A product with high quality and high prices is not better than a similar product with low quality and low price. It depends on the personal preferences and shopping habits to chose any of them. These two products are both good when minimizing the price and maximizing the quality. In this example, however, a cheap high-quality produce is definitely better than an expensive low-quality product.

The mathematical definition of Pareto dominance and Pareto optimality, which is often called Pareto optimality is defined as follows for a minimization problem [2]:

Suppose that there are two vectors such as: $\overrightarrow{x} = (x_1, x_2, ..., x_k)$ and $\overrightarrow{y} = (y_1, y_2, ..., y_k)$. Vector $\overrightarrow{x}$ dominates vector $\overrightarrow{y}$ (denote as $\overrightarrow{x} \prec \overrightarrow{y}$) iff:

$$\forall i \in (1, 2, ..., o)$$

$$[f_i(\overrightarrow{x}) \leq f_i(\overrightarrow{y})] \wedge [\exists i \in 1, 2, ..., o : f_i(\overrightarrow{x}) < f_i(\overrightarrow{y})]$$

A solution $\overrightarrow{x} \in X$ is called Pareto-optimal iff:

$$\{\nexists \overrightarrow{y} \in X | \overrightarrow{y} \prec \overrightarrow{x}\}$$

The mathematical definition of Pareto dominance and Pareto optimality, which is often called Pareto optimality is defined as follows for a maximization problem:

Suppose that there are two vectors such as: $\overrightarrow{x} = (x_1, x_2, ..., x_k)$ and $\overrightarrow{y} = (y_1, y_2, ..., y_k)$. vector $\overrightarrow{x}$ dominates vector $\overrightarrow{y}$ (denote as $\overrightarrow{x} \succ \overrightarrow{y}$) iff:

$$\forall i \in (1, 2, ..., o)$$

$$[f_i(\overrightarrow{x}) \geq f_i(\overrightarrow{y})] \wedge [\exists i \in 1, 2, ..., o : f_i(\overrightarrow{x}) > f_i(\overrightarrow{y})]$$

A solution $\overrightarrow{x} \in X$ is called Pareto-optimal iff:

$$\{\nexists \overrightarrow{y} \in X | \overrightarrow{y} \succ \overrightarrow{x}\}$$

In the example above, there might be thousands and thousands of products with different quality and price that are all good choices for a shopper. The set of all non-dominated solutions for a given problem is called Pareto optimal solution set. This set includes the solutions that represents the best trade-offs between the objectives. The Pareto optimal set is defined as follows:

The set of all Pareto-optimal solutions for a minimization problem:

$$PS := \{\overrightarrow{x}, \overrightarrow{y} \in X | \nexists \overrightarrow{y} \prec \overrightarrow{x}\}$$

The set of all Pareto-optimal solutions for a maximization problem:

$$PS := \{\overrightarrow{x}, \overrightarrow{y} \in X | \nexists \overrightarrow{y} \succ \overrightarrow{x}\}$$

There is another set that is popular in the field of multi-objective optimization called Pareto optimal front. This set has the same number of solutions as to the Pareto optimal set. However, we store the objective values for all objectives of every solution in this set. In other words. Pareto optimal front is the projection of the Pareto optimal solution when considering the objectives only. This set is defined as follows:

$$\forall i \in (1, 2, ..., o)$$

$$PF := \{f_i(\overrightarrow{x}) | \overrightarrow{x} \in PS\}$$

To visually see how these definitions work, Fig. 2.1 is given. This figure shows the Pareto optimal fronts for a bi-objective problem. Four possible cases in which the two objective can be minimized or maximized are consider. In each case the solutions highlighted in the black regions dominate other solutions in the white areas since they show equal objective value in both objectives and better in at least one of them.

2.3   A Classification *f* Multi-objective Optimization Algorithms

There are different classifications for multi-objective optimization algorithms. Since the focus of this book is on optimization using metaheuristics and evolutionary algorithms, the most well-regarded classification in this area is discussed in this section.

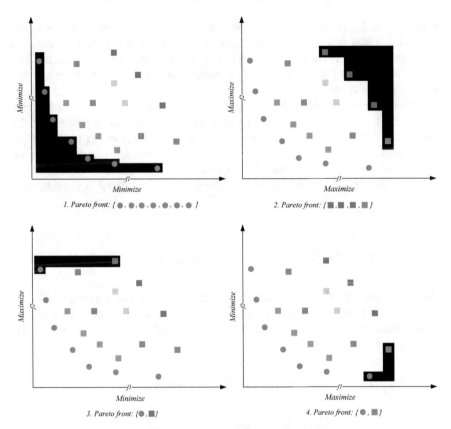

Fig. 2.1 Pareto optimal fronts for a bi-objective problem. Four possible cases in which the two objective can be minimized or maximized are considered. In each case the solutions highlighted in the black regions dominate other solutions in the white areas since they show equal objective value in both objectives and better in at least one of them

This classification is based on the time when a decision maker is involved. This is because from a decision maker perspective, one of the Pareto optimal solutions will be required for a given application. Decision making can be done before, during, or after a multi-objective optimization process. The three classes are: a priori, interactive, and a posterior methods.

In the first class [4], a decision maker is involved before starting the multi-objective optimization process. Therefore, we know how much of each objective is important to the decision maker. This will assist to narrow down the search process to the areas of the search space and objective space that the decision maker is interested in. Such methods are called a priori because we know the importance of objectives prior to the commencement of the optimization process.

In the second class [5], a decision maker is involved during the multi-objective optimization process. This type of optimization is often called human-in-the-loop optimization. In such methods, the optimization process is periodically paused and a decision maker is required to choose the most desirable or promising solution(s) obtained so far. After feeding the algorithm with the preferred solutions, depending on the mechanism the algorithm with consider those solutions in the rest of the multi-objective optimization process.

In the last class [3], decision making is done after the optimization process. This means that the algorithm finds as many Pareto optimal solutions as possible for the problem. A decision maker the decides which solution is good for which applications. Due to the used of a decision maker after the optimization process, such methods are often called a posteriori multi-objective optimization algorithms.

These three types of multi-objective optimization are covered in the following sections.

2.4   A Priori Multi-objective Optimization

One of the most popular a priori methods is aggregation methods, in which multiple objectives are combined into a single objective using a set of weights. This kind of optimization can be formulated as a minimization problem as follows without the loss of generality:

$$Minimize: \quad f(\overrightarrow{x}) = \sum_{i=1}^{o} w_i f_i(\overrightarrow{x}) \tag{2.17}$$

$$Subject\ to: \quad g_i(\overrightarrow{x}) \geq 0, i = 1, 2, ..., m \tag{2.18}$$

$$h_i(\overrightarrow{x}) = 0, i = 1, 2, ..., p \tag{2.19}$$

$$lb_i \leq x_i \leq ub_i, i = 1, 2, ..., n \tag{2.20}$$

where $\overrightarrow{x}$ is a vector that stores all the variables ($\overrightarrow{x} = \{x_1, x_2, x_3, ..., x_{n-1}, x_n\}$) for the problem, n is number of variables, m is the number of inequality constraints, p is the number of equality constraints, lb_i is the lower bound of the i-th variable, and ub_i is the upper bound of the i-th variable.

It can be seen in the above equations that the multiple objectives are aggregated using a set of weights. These ways are defined prior to the optimization process, which is where the name a priori multi-objective optimization comes from. A decision maker shows how much each of the objectives are important with the set of weights. The main advantage of this method is its simplicity and low computational cost since a single-objective algorithm can optimize the aggregated objectives without the need to store and handle non-dominated solutions.

However, there are some drawbacks with a priori methods [6]. Firstly, the set of weight is required that might not be available if there is no decision maker. Secondly,

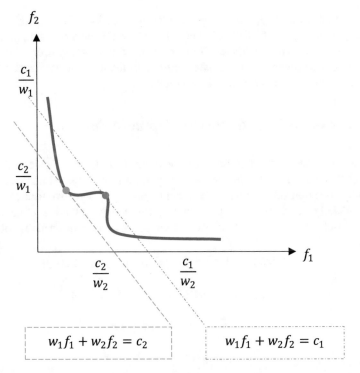

Fig. 2.2 We would like c to be as small as possible because it will lead to a smaller target value (aggregated objective). Figure 2.2 shows the intersection of the straight line and the $X(f_1)$ axis is $\frac{c}{w_1}$, and the intersection of the straight line and the $Y(f_2)$ axis is $\frac{c}{w_2}$. Therefore, smaller c values are obtained the straight line gets close to the origin. Therefore, there is always a point in the convex region that gives smaller c and considered to be a better solution in the non-convex regions. This is the main drawback of aggregation-based methods

the algorithm needs to be run multiple times to find multiple Pareto optimal solutions. Thirdly, to find the Pareto optimal solution set, the weights should be changed. In this case, even changing the weights uniformly does not guarantee finding uniformly distributed Pareto optimal solution. Finally, the non-convex regions of the Pareto optimal front cannot be obtained due to the use of positive weights. This can be seen in Fig. 2.2.

In this figure it is assumed that we want to minimize both objectives. In the aggregation-based method, we have to use the $w_1 * f_1 + w_2 * f_2$ equation to find the Pareto optimal solution. This means that we want to find the intersection of the straight line $w_1 * f_1 + w_2 * f_2 = c$ and the feasible domain.

Obviously, we would like c to be as small as possible because it will lead to a smaller target value (aggregated objective). Figure 2.2 shows the intersection of the straight line and the $X(f_1)$ axis is $\frac{c}{w_1}$, and the intersection of the straight line and the $Y(f_2)$ axis is $\frac{c}{w_2}$. Therefore, smaller c values are obtained the straight line gets close to the origin. Therefore, there is always a point in the convex region that gives smaller c and considered to be a better solution in the non-convex regions. This is the main drawback of aggregation-based methods.

2.5   A Posteriori Multi-objective Optimization

In a posteriori multi-objective optimization algorithm, the multi-objective formulation of the problem is maintained and they are all optimized simultaneously. A multi-objective optimization when maintaining the multi-objective formulation can be formulated as a minimization problem as follows without the loss of generality.

A multi-objective optimization can be formulated as a maximization problem as follows:

$$Minimize: \quad \overrightarrow{F(\vec{x})} = \{f_1(\vec{x}), f_2(\vec{x}), ..., f_o(\vec{x})\} \tag{2.21}$$

$$Subject\ to: \quad g_i(\vec{x}) \geq 0, i = 1, 2, ..., m \tag{2.22}$$

$$h_i(\vec{x}) = 0, i = 1, 2, ..., p \tag{2.23}$$

$$lb_i \leq x_i \leq ub_i, i = 1, 2, ..., n \tag{2.24}$$

where $\vec{x}$ is a vector that stores all the variables ($\vec{x} = \{x_1, x_2, x_3, ..., x_{n-1}, x_n\}$) for the problem, n is number of variables, m is the number of inequality constraints, p is the number of equality constraints, lb_i is the lower bound of the i-th variable, and ub_i is the upper bound of the i-th variable.

It may be see in this formulation that we have a vector of objectives that all should be optimized. In a posteriori method Pareto optimal dominance is used to compare solutions. During the optimization process, therefore, a posteriori optimization algorithm need to store non-dominated solutions as the best solutions for the problem. There are two ultimate goals here. For one, we have to find an accurate approximation of the true Pareto optimal solutions, which is called convergence.

For another, the distribution of solutions across all objectives should be as uniform as possible, which is called coverage. The reason why coverage is important is because in a posteriori method, decision making is done after the optimization process. Therefore, a uniformly distributed Pareto optimal solutions give the decision maker a large number of different solutions to choose from for different applications and purposes.

2.6   Interactive Multi-objective Optimization

In the interactive multi-objective optimization, decision making is done during the optimization process [7]. This means that human's input is fetched to guide the search process to the regions that are of the interests of decision makers. This is why such methods are often called human-in-the-loop optimization.

In interactive optimization methods, users (including decision makers) interact with the optimization algorithm to improve the efficiency of the algorithm, enrich the optimization model, or evaluating the quality of the solution(s) obtained. In interactive multi-objective optimization, a set of random solution is first generated. The user then evaluations some or all of those solutions, provide his/her preferences. The optimization process is then continued with the preferences to find desirable solutions. This process is continued until the user confirms one or multiple of the solutions.

Interactive methods tend to be more efficient than a priori algorithms. This is because a priori method required the preferences to be defined before the optimization process, so there is no opportunity for the decision maker to adjust, adapt, or correct those preferences. In case of any changes in the weights, the whole optimization process should be re-started.

Interactive methods also have several advantages as compared to a posteriori multi-objective optimization algorithm. The set of Pareto optimal solutions used and improved in each iteration in interactive methods is smaller than that in a posteriori method. Therefore, the algorithm does not waste computational resources searching in non-promising regions of the search space.

2.7   Conclusion

This section provided preliminaries and essential definitions in the field of multi-objective optimization. The three classes of multi-objective optimization algorithms were covered as well including: a priori, a posteriori, and interactive methods.

References

1. Deb K (2014) Multi-objective optimization. In: Search methodologies. Springer, Boston, pp 403–449
2. Censor Y (1977) Pareto optimality in multiobjective problems. Appl Math Optim 4(1):41–59
3. Jaszkiewicz A, Branke J (2008) Interactive multiobjective evolutionary algorithms. In: Multi-objective optimization. Springer, Berlin, pp 179–193
4. Marler RT, Arora JS (2010) The weighted sum method for multi-objective optimization: new insights. Struct Multidiscip Optim 41(6):853–862
5. Coello CAC, Lamont GB, Van Veldhuizen DA (2007) Evolutionary algorithms for solving multi-objective problems, vol 5. Springer, New York, pp 79–104

6. Jin Y, Olhofer M, Sendhoff B (2001) Dynamic weighted aggregation for evolutionary multi-objective optimization: why does it work and how? In: Proceedings of the 3rd annual conference on genetic and evolutionary computation. Morgan Kaufmann Publishers Inc, pp 1042–1049
7. Meignan D, Knust S, Frayret JM, Pesant G, Gaud N (2015) A review and taxonomy of inter-active optimization methods in operations research. ACM Trans Inter Intell Syst (TiiS) 5(3):17

Chapter 3
Multi-objective Particle Swarm Optimization

3.1 Introduction

Swarm Intelligence (SI) refers to the collective behaviour of a group of creatures without a centralized unit control. This field was first established in 1989 in a robotic project [1]. Systems built based on SI typically have independent intelligent agents that interact locally to achieve a goal as a team [2]. Most of the algorithms in this field mimic swarm intelligence in nature. For instance, Ant Colony Optimization (ACO) [3] mimics swarm intelligence of ants in an ant colony using stigmergy, which is the communication between individuals in a swarm by modifying environment. It has been proved that ants can find the shortest path between multiple path to a food course from their nest by a depositing and marking the ground using pheromone.

Another evident example of swarm intelligence in nature can be seen in school of fish or flock of birds. In both of these species, there is no centralized control unit to lead the swarm, but local interaction cascade a new direction or predator avoidance throughout the swarm. According to Reynolds [4], there are three rules that can be found largely in such swarms that cause such social and swarm intelligence: alignment, cohesion, and separation. These three behaviors are shown in Fig. 3.1.

This figure shows that the alignment causes each individual to align its flying path towards the average direction of neigbbouring flock mates. Cohesion causes movement towards the average position of neigbbouring flock mates. Finally, the septation prevents individuals from collisions. These simple rules lead to complicated maneuvering mechanism when foraging food or avoiding predators. And example of just the movement can be seen in Fig. 3.2.

In a swarm, the intelligence of each individual causes making the best decision using local information. Such decisions impact other neighbouring individuals as well. This leads to problem solving in swarms as well. For instance, when some individuals on the edge of a swarm finds a food source, they can pull the entire swarms towards it by simply swimming in the direction or around the food. On the

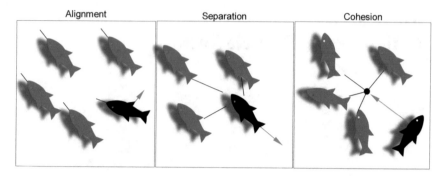

Fig. 3.1 In a school of fish, alignment causes each individual to align its flying path towards the average direction of neigbbouring flock mates. Cohesion causes movement towards the average position of neigbbouring flock mates. Finally, the septation prevents individuals from collisions

Fig. 3.2 Simulation of 50 boids using Raynold rules. I can be seen that there are groups of fishes in each area that alight their movement direction while keeping a safe distance to maintain cohesion and prevent collision

other hand, individual on the edge of swarm can warn and cause predator avoidance across the entire swarm.

3.2 Particle Swarm Optimization

Particle Swarm Optimization (PSO) [5] is one of the most popular swarm intelligence techniques that mimic the navigation mechanism of a swarm of birds of a school of fishes in nature. In this algorithm, a set of random solutions is created first. Each solution is represented with a position vector called $\overrightarrow{X}$. The length of this vector is equal to the number of variables of the problem. The PSO algorithm changes this position vector through a set of iterations until the satisfaction of and end criterion.

The reason why each solution in PSO is called a particle is that we assume solutions can move in an n-dimensional search space where n is the number of variables of the problem, which is often called dimension. The changes in the position vector is done using another vector called the velocity vector ($\vec{V}$). The equations proposed for this algorithm are as follows [6]:

$$\vec{X_i} = \vec{X_i} + \vec{V_i} \tag{3.1}$$

where i is an index that refers to the ith particle.

It was mentioned above that the PSO algorithm changes the position vector of particles several times, this can be shown in the mathematical model as follows [6]:

$$\overrightarrow{X_i(t+1)} = \vec{X_i}(t) + \vec{V_i}(t+1) \tag{3.2}$$

where t shows the number of iteration.

The above equation shows that the next position of a particle is calculating by adding its The main mechanism of PSO is in its velocity updating procedure. It was argued by the inventor of this algorithm that the intelligence of a bird swarm can be simplified into four rules:

• Each individual in the swarm can memorize its best solution obtained so far
• Each individual in the swarm tends to search around its best solution obtained so far
• Each individual in the swarm can see the best solution obtained by the entire swarm at any given time
• Each individual in the swarm gravitates towards the best solution found by the swarm

Note that the first two rules mimic the social individual intelligence of a an individual in a swarm, which is often called cognitive intelligence. The last four rules simulated the social interaction between the individuals in the swarm called social intelligence. The equation to update the velocity in PSO is as follows:

$$\overrightarrow{V_i(t+1)} = w\vec{V_i}(t) + c_1 r_1(\overrightarrow{P_i(t)} - \vec{X_i}(t)) + c_2 r_2(\vec{G}(t) - \vec{X_i}(t)) \tag{3.3}$$

where w shows the inertial weight, r_1 is a random number in [0, 1], r_2 is a random number in [0, 1], c_1 is a coefficient to tune the impact of the social component, c_2 is a coefficient to tune the impact of the cognitive component, $P_i(t)$ indicates the position of the best solution obtained by the ith particle, and $G(t)$ is the position of the best solution found by the entire swarm at tth iteration.

The velocity updating equation is made of three component. The firs component, $w\vec{V_i}(t)$ maintains a portion of the current speeds. The inertial weight indicates how much the particle keeps its momentum. Due to the accumulative nature of the position updating equation in PSO, the inertial weight should be set to a number less than 1. Otherwise, the particles goes outsides the search area that they are supposed to search.

The second component simulates the individual intelligence of each particle. The best solution obtained by each particle is stored in $\overrightarrow{P_i(t)}$, and its distance from the current position for the particle indicates how much the particle should gravitates towards its personal best. The magnitude of this tendency is defined using c_1 and r_1. The parameter c_1 is normally set to 2, and r_1 is a random number in [0, 1] generated from a Gaussian distribution. This part of the velocity updating equation shows stochastic behaviour due to the use of a random number that is typically generated for each particle and each dimension.

The last component of the velocity equation considers the distance between the current position of the particle and the position of the best solution obtained by the PSO algorithm. A similar stochastic component to the cognitive component is used to provide random behaviour when moving towards the best solution obtained by the swarm. There are two parameters called c_2 and r_2 where c_2 indicates the impact of the social intelligence and r_2 is a random number in [0, 1] generated from a Gaussian distribution.

To show how the mechanism of PSO works when solving optimization problems. The search history of particles when solving the function shown in Fig. 3.3. This function has a large number of local solutions and the formulation ($f(x) = \sum_{i=1}^{n}(x^2 - 10cos(2\pi x_i)) + 10n$) allows increasing the number of variables.

A PSO algorithm with 10 particles and 100 iterations is used to solve a 3-dimensional version of the test function. The interial weight is linearly decreased from 0.9 to 0.2. Both cognitive and social coefficient are set to 1. The results of one run is given in Fig. 3.4. Note that the points in the initial search are visualize with cool colors. The colors gradually change to warm colors proportional to the number of iteration.

This figure shows that the PSO algorithm covers a descent potion of the search space and eventually converges to a solution. The subplots around the main plot show the position of particles in a 2D space when considering each pair of the three variables. The 2D projections also show that the search is directed and not

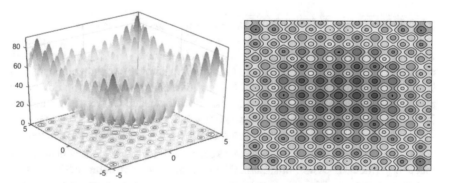

Fig. 3.3 Rastrigin function used to test the performance of PSO: $f(x) = \sum_{i=1}^{n}(x^2 - 10cos(2\pi x_i)) + 10n$

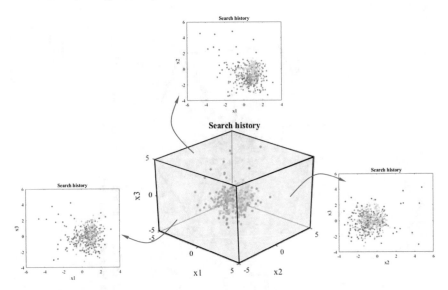

Fig. 3.4 The points in the initial search are visualize with cool colors. The colors gradually chage to warm colors proportional to the number of iteration. Good coverage of the search space and convergence of particles can be observed. The subplots around the main plot show the position of particles in a 2D space when considering each pair of the three variables. The 2D projections also show that the search is directed and not purely random

purely random. This is due to the stochastic nature of PSO, in which particles move randomly around the search space. However, such random movements are guided by personal best solutions and the swarm best solution. Also, the inertial weight linearly decreases, which means particles maintain lesser of their momentum as the iteration counter increases. Tuning the controlling parameters of PSO (w, c_1, and c_2) will results in different search patterns.

The key point here is that the PSO algorithm is not a complete search and should be used when we cannot afford using a complete search like a brute force search. Of course, for computationally cheap problems a complete search method is a better choice. On other hard, a pure random search can be used to search for the global optimum of an expensive problem. However, the search is not guided at all. To observe the difference between these three methods, Fig. 3.5 is given.

Figure 3.5 shows that a complete search provide a very regular search pattern and it is a guarantee to find the global solution. Note that in this experiment we used discrete values for the brute force search for the sake of visualization. In reality, the grid shape in Fig. 3.5 will be a solid cube since all points inside it should be searched. The complexity of the complete search is also of $O(n^3)$ where n is the number of discrete values chosen for each axis. The second subplot shows that a pure random search show a good coverage of the search space. However, this search is not directed at all. The complexity of the search in the worst case is $O(\infty)$, which mean such as search might take forever to find the solution.

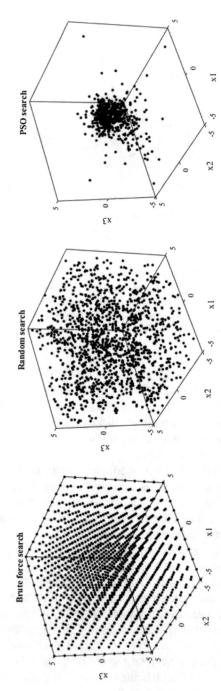

Fig. 3.5 (left) a complete search provide a very regular search pattern and it is a guarantee to find the global solution (middle) a pure random search show a good coverage of the search space (right) The PSO search pattern. The search is not as broad as the other two methods, but it is guided

The last subplot in Fig. 3.5 shows that the search is not as broad as the other two methods, but it is guided. In fact, the most promising regions of the search space is searched using the personal bests and the swarm best solutions. The computational complexity of the PSO algorithm is of $O(dnt)$ where d is the number of variables, n shows the number of particles, and t indicates the number of iterations. This makes is less computationally expensive than the brute force search which comes at the cost of not being a complete search.

In PSO, a set of parameters is first set including: number of particles, stopping condition, cognitive constant, social constant, and inertial weight. The algorithm then creates a set of random particles in an n-dimensional search space. Until the satisfaction of the end criterion, the algorithm constantly updating the inertial weight, velocity vectors, positions vectors, evaluation the particles using the objective function, the personal bests, and the global best. Once the end criterion is satisfied, the algorithm returns the global best as the best approximation of the global optimum for the problem.

The PSO algorithm is good to solve single-objective problems. This is due to the use of relational operators to compare solutions against the personal best and the global best. In a multi-objective problem, however, there is more than one criterion to compare. In order to solve such problem using PSO, the easiest way is to combine the objectives into a single objective. However, this method has several drawbacks that was discussed before. To maintain the multi-objective formulation and solve it using PSO, there has no be changes in the mechanisms of this algorithm.

3.3 Multi-objective Particle Swarm Optimization

In the PSO algorithm, solutions are not compared together. Each particle is an independent that use the position and velocity vectors to move in the search space. The only way that the particles communicate is via the best solution obtained so far. This is allows the particles to individual search different promising regions of the search space indicated by the personal best solution found so far while leaning towards to best solution obtained by the entire swarm. Therefore, each particle is able to lead the entire swarm towards a region of the search space if its personal best is better than everything else. There are other topologies in PSO as well such as fully-connected, von Neumann, and random [7, 8]. Each of these topologies impact the exploration and exploitation of the PSO algorithm. For the sake of simplicity, this books considers the first topology for all the PSO algorithms.

The global best topology requires comparison in two places: comparison with the personal best and comparison with the global best. In both case, we cannot use relational operators when solving multi-objective problems. In the MOPSO algorithm, it is assumed that we have to use Pareto optimality dominance operator to perform comparison. In the most popular MOPSO version in the literature, Coello and Lechuga [9] designed and considered a repository just like the Pareto Archived Evolution Strategy (PEAS) [10] to store the non-dominated solutions obtained by the particles. This repository is often called archive. This archive saves all the best solutions obtained so say by the swarm, so we are in fact dealing with multiple global best solution in each iteration. To search as broad as possible, MOPSO selects different solutions from the archive when updating the position of particles independently. The particles do not need any archive to store all their personal best solution. So they are still searching promising regions of the search space, but at the same time gravitates towards different non-dominated solutions in the archive. This will lead to a balance between local and global search in PSO.

Using archive might lead to two issues that should be handled in MOPSO. Firstly, the archive size grows over time and the number of dominated solutions significantly increases when the number of objectives increases in a problem. Therefore, the size of archive should be limited to avoid increasing the run time of searching and updating non-dominated solutions in the archive. Secondly, the dominated solutions in the archive should be found of replaced with non-dominated solutions by the particles. Sometimes all the solutions are non-dominated in the archive and there are still other non-dominated solutions found. Due to the limited size of archive we have to make a decision at some point which non-dominated solution to keep and which one to remove. Thirdly, there should rules on how to choose non-dominated solutions in the archive as the global best for each particle. To resolved all these issues the followings are considered and implemented into the MOPSO algorithms.

To avoid high computation cost of searching and maintaining the archive, it's size is assumed fixed. Therefore, the MOPSO algorithm is able to find a maximum of n non-dominated solutions where n is defined as a parameter in MOPSO. Once the archive becomes full, then we have to organize it in a way to accommodate better non-dominated solutions. The archive has been equipped with a grid mechanism that divide the objective space of the problem into a grid. To increase the coverage of the Pareto optimal solution set and front, non-dominated solutions from crowded regions of the grid should be removed. Instead, they are replaced with non-dominated solutions that lie on the less crowded areas of the archive. This method allows MOPSO to constantly improve the distributions of non-dominated in the archive across all the objectives. On the other hand, the global best for each particle should be chosen from the least crowded area to again increase the coverage of solutions. This is shown in Fig. 3.6.

To resolve the last issue on how to add non-dominated solutions in the archive, the MOPSO algorithm follows these rules [9]:

- If the archive is empty and there is a particle that is non-dominated, it should be added to the archive.
- If a solution in the archive is dominated with respect to a solution outside the archive (a particle), it should be replaced with the new solutions immediately.
- If a solution is non-dominated in comparison with the solutions in the archive and we have enough space, the solution should be added to the archive.
- If a solution is non-dominated in comparison with the solutions in the archive and we have do not have enough space, on solution in the most crowded segments of

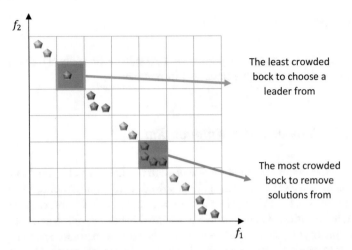

Fig. 3.6 To increase the coverage of the Pareto optimal solution set and front, non-dominated solutions from crowded regions of the grid should be removed. Instead, they are replaced with non-dominated solutions that lie on the less crowded areas of the archive. This method allows MOPSO to constantly improve the distributions of non-dominated in the archive across all the objectives. On the other hand, the global best for each particle should be chosen from the least crowded area to again increase the coverage of solutions

the archive grid should be removed and the new solution should be inserted to the archive.

It should be noted that process of removing a solution from a crowded region and choosing a global best from a less crowded region should be done based on a probability to avoid local solutions in the search space too. This is done by counting the number of solutions in each segment of the grid and using it as normal or inverse to calculate the probability. The more the number of non-dominated in a segment, the higher probability of removing one solution and the less probability of choosing a global best from.

The mesh in MOPSO is made of hypercube. The number of dimensions is equal to the number of objectives. The process of updating the archive can be of $O(n^2)$ in the worst case. This makes it computationally cheaper than niching methods as of one the most popular techniques to improve the coverage of Pareto optimal solutions in the Pareto optimal solution set.

The investors of MOPSO argued that the MOPSO algorithm might not show high exploratory behaviour. Therefore, they introduced a mutation operator that changes the particles randomly using the upper and lower bounds of their variables. The mutation operator is desirable at the beginning of the optimization, but its effect should be faded away as the algorithm gets closer to the final iterations. This is done in the original MOPSO algorithm by decreasing the number of particles that are mutated.

3.4 Results

In this section, a wide range of experiments are conducted to analyze the performance of MOPSO and investigate the impact of its parameters.

3.4.1 The Impact of the Mutation Rate

In the first experiment the mutation rate of MOPSO has been changed to observe its performance. The ZDT1 problem [11] is solved using 100 particles, 30 iterations, archive size of 100, inertial weight of 0.5, $c1 = 1$, and $c2 = 2$. To observe the impact of mutation, the position of each particle is recorded in each iteration and visualized as the search history. To better observe patterns, the color of particles are changed from cool colors to warm colors proportional to the number of iterations. The reason why more particle is chosen and only 30 iteration is to avoid overlapping many points in the last iterations. As the particles get closer the to Pareto optimal front, their movements get smaller so the points in the figure will overlap and it will be difficult to observe any patters. Also, the main purpose of this experiment is to see how the mutation rate change the search pattern of MOPSO and not necessarily how it finds an accurate estimation of the true Pareto optimal solution. Four mutation rates are used in Fig. 3.7 0, 0.1, 0.5. and 0.9.

Random changes in particles increase proportional to the value of the mutation rate. As expected, Fig. 3.7 shows that when the mutation rate is equal to 0, the search history of particle is sparse. Most of the particles quickly coverages towards the Pareto optimal solution (red regions). This is beneficial for the test function since there is no local solutions and fronts. In problems with locally optimal solutions and Pareto optimal front, however, the lack of mutation is likely to results in local optima (fronts) stagnation. The second subplot in Fig. 3.7 shows that the search history is less sparse when the mutation rate is equal to 0.1. This is due to random changes in some of the particles in each iteration. Despite the less accurate of the front obtained as can be seen in the red regions of the second subplot, such behaviour is fruitful for problems with local solutions since the algorithm need to avoid them and explore the search space as extensively as possible at the beginning of the optimization process.

The last two subplots (c, d) in Fig. 3.7 show that the search pattern of MOPSO can be change significantly if the mutation rate is set to large values. In subplot (c), the bottom-left part of the Pareto optimal front is not found with a good distribution. This is because a large number of particles face random changes and do not follow the entire swarm towards the most promising regions obtained so far. The subplot (b) shows the least convergence towards the true Pareto optimal front because 90% of times a particle is likely to be mutation. So, the search is more random than direct. This experiment shows that the mutation mechanism is beneficial in MOPSO, but it should be tuned carefully to avoid damaging the search patter of the swarm in MOPSO. As a general advise, it is recommended to set it below 0.5. It is also

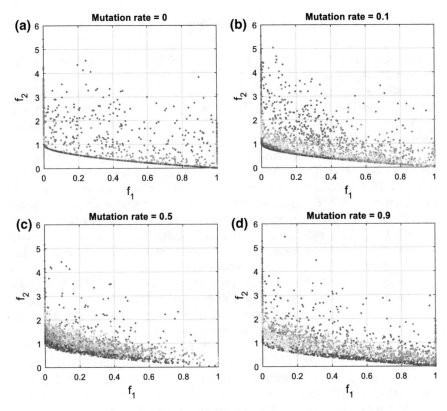

Fig. 3.7 When the mutation rate is equal to 0, the search history of particle is sparse. As a consequence, Most of the particles quickly coverages towards the Pareto optimal solution (red regions), which is beneficial for the test function since there is no local solutions and fronts. In problems with locally optimal solutions and Pareto optimal front, however, the lack of mutation is likely to results in local optima (fronts) stagnation. The coverage of space and exploratory behaviour of MOPSO increases proportional to the mutation rate

recommended to tune this parameters for the problem on hand side the performance MOPSO (as a stochastic algorithm) varies when changing the problem.

It is worth mentioning here that the impact of mutation on MOPSO and PSO similar to that in MOGA and GA. In GA, the crossover operator causes gene exchange between individuals in a population. This results is exploitation of the "areas" between two or multiple genes depending on the type of crossover. Mutation, however, introduce new genes with randomly changing them which results in searching the areas that are outside the reach of the genes.

3.4.2   The Impact of the Inertial Weight

The inventors of the MOPSO algorithm claimed that the main mechanism of exploration in this algorithm is mutation and the inertial weight does not have to be tuned. To investigate this, this subsection tune this parameter and observe its impact on the search history of particles on the objective space. The ZDT problem is solved using 100 particles, 30 iterations, archive size of 100, c1 $= 2$, and c2 $= 2$. The search history of particles are given in Fig. 3.8.

The results in subplot (a) shows that the search history if very sparse. There is no continuity between the solutions far from the Pareto optimal front and those very close to it. This is because when the inertial is high, particles maintaining their speed and direction to update their position. This reduces the impact of social and cognitive components significantly. So particles tend to be highly independent when the inertial wright is high. The subplot (b) shows that the search pattern changes slightly, and there are green and yellow points between the initial blue and final red points. This shows more information exchange between particles.

The subplots (c) and (d) show that the coverage of solutions is slightly better than the subplot (a) when the inertial weight is equal to 0.9. The search pattern changes more when there is the inertial weight is equal to 0. This is because in this case, particles only update position based on their social and cognitive components. The particles doe not consider their current velocity at all, which results in less directed changes. Just like PSO, the most ideal case is to linearly change the inertial weight. The last subplot in Fig. 3.8 shows that the continuity between the sampled point is highly uniform and the convergence is also good. Therefore, these results show that it is better to use time-varying values for the inertial weight.

Overall, the results in Fig. 3.8 confirm that the inertial weight does not change the search pattern significantly and it seem that high values for the inertial wright results in more exploitation than exploration in the objective space.

3.4.3   The Impact of Personal (c_1) and Social (c_2) Coefficients

In this section, the impact of two parameters called personal (c_1) and social (c_2) coefficients on the performance of MOPSO is investigated. The ZDT problem is solved using 100 particles, 30 iterations, archive size of 100, and inertial weight of 0.5. The search history of particles are given in Fig. 3.9.

The subplot (a) shows that when both personal and social components are set to zero, the MOPSO algorithm does not converge at all. It can be seen that each particle moves slightly and finds a solution in the search space. However, the search is not systematic and the solutions found by each particles are far from the Pareto optimal solutions. In subplot (b) the social component is equal to zero, which means that the particles only search around their own personal bests. There is no interaction between particles, so this is why multiple traces can be seen in subplot (b) that

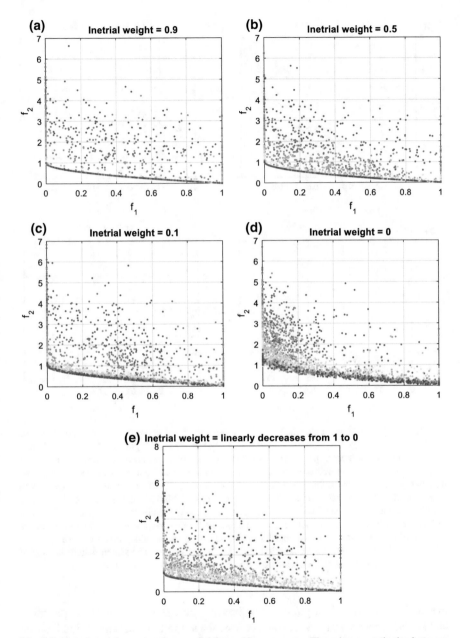

Fig. 3.8 Subplot **a** shows that the search history if very sparse. There is no continuity between the solutions far from the Pareto optimal front and those very close to it. Subplot **b** shows that the search pattern changes slightly, and there are green and yellow points between the initial blue and final red points. This shows more information exchange between particles. The subplots **c** and **d** show that the coverage of solutions is slightly better than the subplot (**a**) when the inertial weight is equal to 0.9. The search pattern changes more when there is the inertial weight is equal to 0

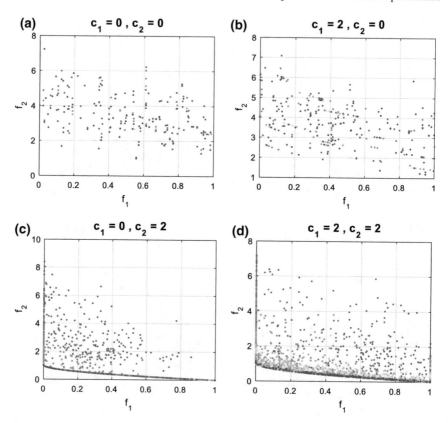

Fig. 3.9 Subplot **a** shows that when both personal and social components are set to zero, the MOPSO algorithm does not converge at all. It can be seen that each particle moves slightly and finds a solution in the search space. In subplot **b** the social component is equal to zero, which means that the particles only search around their own personal bests. There is no interaction between particles, so this is why multiple traces can be seen in subplot (**b**) that eventually converges to a point in the search space. Subplot **c** shows the results of MOPSO when there is no personal intelligence and all particles just follows the best solution obtained by the entire swarm. It can be seen that the search process is systematic in this case and convergence is evident. Subplot **d** shows that when both personal and social coefficients are set to a positive value other than zero (2 in this case), a good balance between exploration and exploitation occurs. The search pattern in subplot (**d**) smoothly guide the particles towards the final Pareto optimal front obtained

eventually converges to a point in the search space. Despite a higher exploration and slightly more systematic search process, the algorithm still fails to find the true Pareto optimal solutions and Pareto optimal front.

As opposed to subplot (c), the subplot (c) shows the results of MOPSO when there is no personal intelligence and all particles just follows the best solution obtained by the entire swarm. It can be seen that the search process is systematic in this case and convergence is evident. However, the algorithm is prone to local front (or optima) stagnations. The current ZDT test function has only one global front without any local fronts. Therefore, such search mechanisms is beneficial. In a search space with many local solutions and fronts, the algorithm is at the risk of being trapped in local solutions.

The last subplot in Fig. 3.9 shows that when both personal and social coefficients are set to a positive value other than zero (2 in this case), a good balance between exploration and exploitation occurs. The search pattern in subplot (d) smoothly guide the particles towards the final Pareto optimal front obtained.

3.5 Conclusion

In this chapter the MOPSO algorithm was presented as one of the most well-regarded a posteriori algorithms in the field of swarm intelligence. It was discussed that this algorithm has been equipped with an archive and an archive controller to search for the Pareto optimal solutions of a given multi-objective optimization problem. Since MOPSO uses most of equations in PSO, the PSO algorithm was discussed in details as well.

The chapter also included several experiments to better understand the behavior of this algorithm. The main controlling parameters of MOPSO were tuned to find out their impact on the search pattern fo MOPSO. Also, there were several recommendations on the best values for the parameters investigated.

References

1. Beni G, Wang J (1993) Swarm intelligence in cellular robotic systems. In: Robots and biological systems: towards a new bionics? Springer, Berlin, pp 703–712
2. Kennedy J (2006) Swarm intelligence. In: Handbook of nature-inspired and innovative computing. Springer, Boston, pp 187–219
3. Dorigo M, Di Caro G (1999) Ant colony optimization: a new metaheuristic. In: Proceedings of the 1999 congress on evolutionary computation-CEC99 (Cat. No. 99TH8406), vol 2. IEEE, pp 1470–1477
4. Reynolds CW (1987) Flocks, herds and schools: a distributed behavioral model. In ACM SIGGRAPH computer graphics, vol 21, no 4. ACM, pp 25–34
5. Kennedy J (2010) Particle swarm optimization. In: Encyclopedia of machine learning, pp 760–766
6. Shi Y (2001) Particle swarm optimization: developments, applications and resources. In Proceedings of the 2001 congress on evolutionary computation (IEEE Cat. No. 01TH8546), vol 1. IEEE, pp 81–86
7. Kennedy J (1999) Small worlds and mega-minds: effects of neighborhood topology on particle swarm performance. In: Proceedings of the 1999 congress on evolutionary computation-CEC99 (Cat. No. 99TH8406), vol 3. IEEE, pp 1931–1938
8. Kennedy J, Mendes R (2002) Population structure and particle swarm performance. In: Proceedings of the 2002 congress on evolutionary computation, CEC'02 (Cat. No. 02TH8600), vol 2, pp 1671–1676. IEEE

9. Coello CC, Lechuga MS (2002) MOPSO: a proposal for multiple objective particle swarm optimization. In: Proceedings of the 2002 congress on evolutionary computation, CEC'02 (Cat. No. 02TH8600), vol 2. IEEE, pp 1051–1056
10. Knowles J, Corne D (2003) Properties of an adaptive archiving algorithm for storing nondominated vectors. IEEE Trans Evol Comput 7(2):100–116
11. Zitzler E, Brockhoff D, Thiele L (2007) The hypervolume indicator revisited: on the design of Pareto-compliant indicators via weighted integration. In: International conference on evolutionary multi-criterion optimization. Springer, Berlin, pp 862–876

Chapter 4
Non-dominated Sorting Genetic Algorithm

4.1 Introduction

Evolutionary Algorithms mimic natural evolutionary process in nature. One of the most well-regarded evolutionary algorithms is Genetic Algorithm (GA) [1]. This algorithm has been inspired from the Drawin's theory of evolutionary. This theory states that natural organisms develop using the natural selection. Natural selection refers to the process of selecting and propagating the best genes from one generation to the next that helps the new generation to better survive, compete, and reproduce.

In the GA algorithm, each solution is represented as a vector of variables. This vector is often called chromosome or individual. Each variable in this vector is called a gene. Since GA is a population-based algorithm, multiple chromosomes are creating randomly first and are considered as the first generation. All of the chromosomes are evaluated using an objective function in case of single-objective function. The objective value calculated for each of chromosomes is called fitness value.

To create the next generation, the GA algorithm uses three main operators: selection, reproduction (crossover), and mutation. In the basic model of a selection operator, two chromosomes are selected from the population. This selection is done based on the fitness of chromosomes. The better the fitness value, the higher chance of being selection by the selection operator. A roulette wheel is often used to assign a probability of selection to each individual. This allows having small probability to choose less fit individuals. This will assist GA to increase the diversity of chromosomes and resolve local optimal entrapment.

After selecting two chromosomes (called parents), they have to be combined using the recombination operator to produce new chromosomes (called children) for the next generation. The easiest to do this is to use the single-point crossover, in which an imaginary point is chosen randomly in the chromosomes. The two parts before and after this imaginary wall are exchanged between the parents to create children.

The selection and recombination mix the genes in chromosomes during the generations. However, there is no new gene is generated using them. This might lead to

S. Mirjalili and J. S. Dong, *Multi-Objective Optimization using Artificial Intelligence Techniques*, SpringerBriefs in Computational Intelligence,
https://doi.org/10.1007/978-3-030-24835-2_4

less diversity and finding local optima. In nature, some genes face mutation to help a species to handle a new environmental condition. This is simulated in GA using the mutation operator, in which there is a certain probability that a gene might be replaced by a random value. So, if the GA algorithm faces no improvement in the fitness of chromosomes, there is still a chance to find other better solutions.

The GA algorithm repeatedly generate new populations using the aforementioned three operators until the satisfaction of the an end criterion. Of course, GA is an informed search and not complete. This means that this algorithm uses heuristic information to narrow down the search because a complete search is not affordable for the problem because of time and/or space complexities. There is no guarantee that this algorithm finds the best solution of a given problem. Since we chose the best solution each time, however, this algorithm is likely to find better solutions by combining them. At least, this algorithm is potentially better than a complete random search since it makes "educated decisions".

The GA algorithm cannot solve problems with multiple objectives. This is because the selection operator requires a criterion to judge about the quality of solutions. When we use multiple objectives, there is normally more than one best solution. Therefore, the GA algorithm requires modifications to solve multi-objective problems.

4.2   Multi-objective Genetic Algorithm

There have been several versions of multi-objective GA in the literature. Undoubtedly, the most well-regarded ones have been NSGA and NSGA-II [2]. In both methods GA is modified in a way to rank individuals based on their domination level which considers multiple objectives intrinsically. NSGA-II is better than NSGA in terms of computational cost and elitism.

NSGA-II uses a fast non-dominated sorting technique, an elitist-keeping technique, and a new niching operator which is parameterless. These elements are explained in the following paragraphs [2].

Fast non-dominated sorting: The non-dominated sort of NSGA is in the order of $O(MN^3)$ where N and M are the number of individuals in population and the objective functions respectively. This order is due to the comparison of all particles based on all the objective functions together (MN^2) and sorting in non-dominated levels ($MN^2 * N$). In the new method a hierarchical model of non-dominated levels has been proposed to reduce this computational cost whereby we do not need to compare all the dominated individuals after the first non-dominated level ($O(MN^2)$). There are two counters for each individual which show how many individuals dominate it and how many individuals it dominates. These counters help to build the domination levels.

Elitist-keeping technique: Elitism is automatically done due to the comparison of the current population with the previously found best non-dominated solutions. In the GA algorithm, there is a probability of elitism in which a portion of the current population is maintained and moved to the next generation intact. In multi-objective

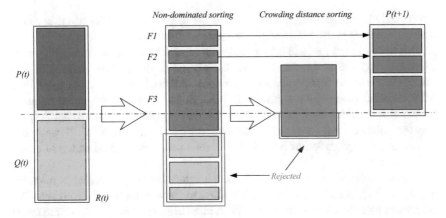

Fig. 4.1 Procedure of NSGA-II [2]

GAs including NSGA-II, this is not implemented due to the existence of multiple best solutions when solving multi-objective problems. However, there are variants of NSGA that uses elitism [3].

New niching operator (crowding-distance [4]): The nearest neighbour density estimates the perimeter of a rectangle (cube or hyper-cube) neighbourhood which is formed by the nearest neighbours. The individuals with higher value of this measures are selected as the leaders. In the niching technique a diameter (σ_{share}) should be defined, and the results highly depends on it. However, there is no parameter to define in the proposed operator.

The procedure of NSGA-II is depicted in Fig. 4.1.

This figure shows how NSGA-II used non-dominated sorting method to sort and rank individuals. The selection operator will then choose any of them based on their probabilities to be participated in the reproduction.

4.3   Results

In this section, several experiments are conducted to analyze the performance of NSGA-II algorithm. The ZDT1, ZDT2, ZDT3 [5], Kursawe [6], Poloni [7] test functions used once more in all the experiments.

4.3.1   The Impact of the Mutation Rate (P_m)

In this experiments, the probability of mutation in NSGA-II is changed to observe its impact of the search behaviour of this algorithm. To find the Pareto optimal front using this algorithm, a population size of 100 and the maximum generation of 100 are chosen. The probability of crossover is set to 0.9. The test functions is ZDT1 with 30 variables is used. The results are presented in Fig. 4.2.

This figure shows that when the probability of mutation is set to 0, individuals are combined only and there is no exploration. The points int he first suplot with cool colors show that the first population is random, so there is a little bit of exploration. After combining individuals, however, they gravitate towards the non-dominated points. The solutions have moved toward the Pareto optimal solutions, but they are very few solutions that are explored (the eight red dots in this example). These results show that the exploration of NSGA-II is at the minimum level when the mutation rate is set to 0.

The next subplot shows that even for small values for the probability of mutation (e.g $P_m = 0.1$), the NSGA-II algorithm starts to show exploratory behaviour. The second subplot shows that there are still streams of lines from the top of the figure to the bottom. However, there are many solutions between each 'steams' as well. This is due to the random changes to the genes with the probability of 10%. It is also interesting that the solutions obtained at the end (red dots) are closer to the Pareto optimal front as compared to the subplot that shows the mutation rate of 0.

The exploration level of NSGA-II is also high when $P_m = 0.2$. This third subplot shows that the solutions are spread well between of the streams. Th convergence of the algorithm is still evident. As the probability of mutation increases, the last subplots show that the exploration becomes more than exploitation. The result to disappearance of the solution streams when $P_m = 0.6$ and $P_m = 0.1$. The trace of solutions from cool color to warm color can be slightly seen when $P_m = 0.4$, but ones this rate gives above 0.5, the individuals move more randomly than in a guided manner. The last two subplots show that he algorithm shows minimum convergence when the probability of mutation is high. When $P_m = 1$, in fact, every gene in every individual face random change which interrupts the process of search in NSGA-II. This is where the exploration becomes completely dominant as compared to exploitation.

Another interesting observation is the high distribution of solutions across all objectives when increasing the probability of mutation. High distribution in NSGA-II is important since it is considered as a posteriori multi-objective optimization algorithm. Although the highest distribution occurs when the probability of mutation is at the maximum, we have to set it to lower values to avoid significant reduction in the exploitation and convergence. Figure 4.2 sows that reasonable convergence and coverage are chived when the probability of mutation is between 0.1 and 0.5. Note that the convergence can be improved with tuning the probability of crossover that will be discussed in the next subsection.

The results of this subsection show that impact of mutation rate on the performance of the NSGA-II algorithm. In most single- and multi-objective evolutionary algorithms, mutation operators are the main mechanism of exploration. This is because they introduce random changes in genes, so if the algorithm gets trapped in local solutions, it has the potential to resolve this issue. The results of this subsection also showed that the mutation rate of NSGA-II better to be set to a value less than or equal to 0.5.

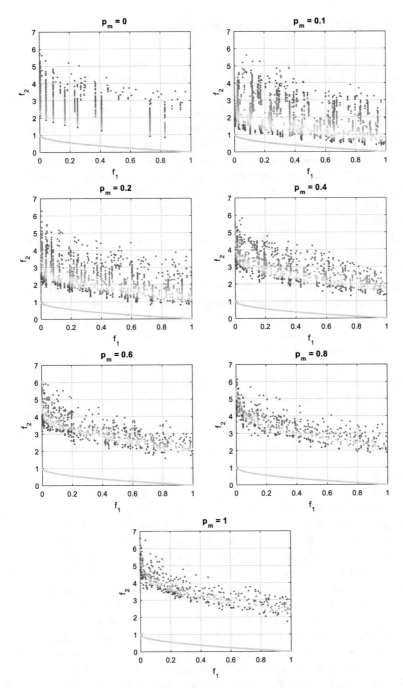

Fig. 4.2 When he probability of mutation is set to 0, individuals are combined only and there is no exploration. For small values for the probability of mutation (e.g $P_m = 0.1$), the NSGA-II algorithm starts to show exploratory behaviour

4.3.2   The Impact of the Crossover Rate (P_c)

In this experiments, the probability of mutation in NSGA-II is changed to observe its impact of the search behaviour of this algorithm. To find the Pareto optimal front using this algorithm, a population size of 100 and the maximum generation of 100 are chosen. The probability of mutation is set to 0. The reason why the value of zero is considered is because we want to have minimum exploration and see if the crossover can compensate it. The test functions is ZDT1 with 30 variables is used. The results are presented in Fig. 4.3. Note that the probability of crossover indicates whether a child or a parent should be moved to the next generation. On one side of the spectrum, no children will be moved into the new population when the probability of crossover is zero. On the other side, all children are moved to the new population when the probability of cross over is 1.

This figure shows that when the probability of crossover is zero (as the probability of mutation), there is no changes in the random population. Figure 4.3 shows that when $P_c = 0.1$, vertical lines can be seen that show individuals are converging towards some non-dominated solutions. Although there is no exploration, chromosomes keep exchanging genes and producing children. Those children are moved to the new population 10% of times in the second subplot. This subplot shows that this cause movement towards the Pareto optimal solutions.

The rest of subplot show that the exploitation and convergence are more evident when we increase the probability of crossover. The closest solutions to the Pareto optimal solutions are found when the probability of crossover is set to 1. Of course, such behaviour is beneficial for both objective of ZDT1 since there do not have any local solutions as shown in Fig. 4.4. For such a problem, there is no need to explore the search space since the slope leads the individuals towards the Pareto Optimal front. For a problem with locally optimal solution, mutation should be incorporated as well.

Another interesting patten is that the distribution of the approximation of the Pareto optimal front is not good in case of the cases. This is again due to the lack of mutation in genes. Random fluctuations in genes results in higher distribution across all objective that will eventually increase the coverage of solutions obtained.

To see how the NSGA-II algorithm performs on problems with different shapes of search space and Pareto optimal front. This algorithm is applied to ZDT2, ZDT3, Kursawe, and Poloni. The NSGA-II was equipped with 100 individuals and 500 maximum number of generations. To have good balance of exploration and exploitation, the probability of crossover and mutation are set to 0.9 and 0.5 respectively. NSGA-II has been run multiple times and the best results are visualized in Fig. 4.5. Note that better results can be obtained with increasing the size of population and the number of generations. However, the main intention of this experiment is to just observe the performance of NSGA-II on different types of problem.

The results in Fig. 4.5 show that the Pareto front of ZDT2 is convex as opposed to the ZDT1. It can be seen that the NSGA-II algorithm is still able to converge and find well distrusted solutions for these types of problems as well. The ZDT3 test function

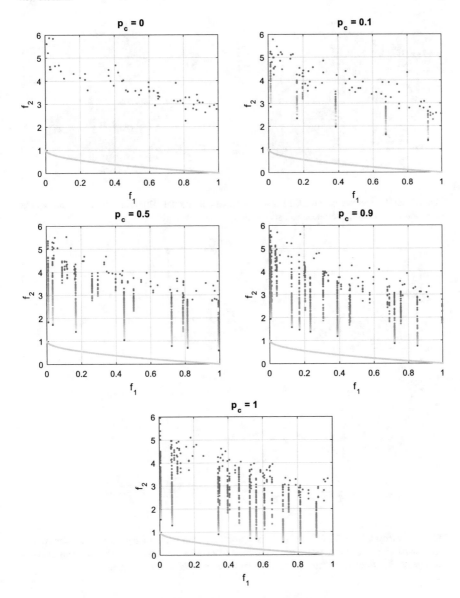

Fig. 4.3 In case of using a fixed (or non-fixed) value for the probability of mutation, the convergence of NSGA-II increases proportional to the value of crossover probability

has Pareto optimal front with separated regions. Therefore, an algorithm needs to find solutions on each five areas to provide best trade-offs between the objectives for decision makers. Figure 4.5 shows that the NSGA-II algorithm performs really well on this problem. The convergence and coverage are both accurate.

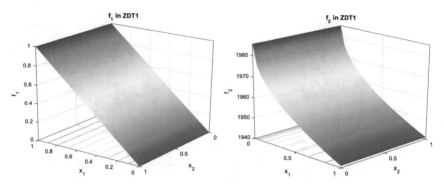

Fig. 4.4 Both objective of the ZDT1 test function are unimodal. Therefore, an algorithm with no exploration but only exploitation can solve them

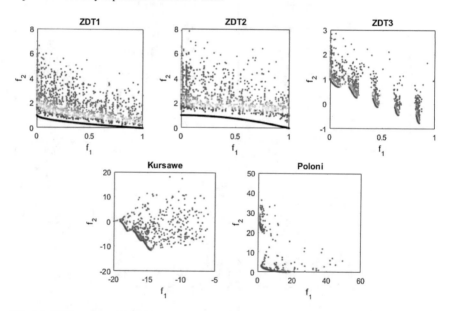

Fig. 4.5 Performance of NSGA-II on other problems with types of Pareto optimal fronts. The Pareto front of ZDT2 is concave. The Pareto front of ZDT2 is convex. The ZDT3 test function has five separated regions. The Kursawe and Poloni have both two separated regions

The Pareto fronts of the last two problems, Kursawe and Poloni, have separated regions. It can be observed that the NSGA-II algorithm finds an accurate estimation of the true Pareto optimal front for these problems as well. What makes the results of NSGA-II different here is the fact that there is no smooth transition between the initial random solutions and the final solutions. It seems that the algorithm quickly finds very close solutions to the Pareto optimal solutions and then start searching around them. This is due to the low-dimensionality of the last two problems. Where there are less number of variables, the search space is very small and it is easy to find reasonably good non-dominated solutions.

4.3.3 Conclusion

This chapter presented the NSGA-II algorithm as the most well-regarded posteriori evolutionary multi-objective optimization algorithm. After discussing the structure of this algorithm, several experiments were conducted to analyze the impact of the main controlling parameters on the performance of NSGA-II. Based on the observations, several recommendations made no how to efficiently tune the parameters of this algorithm.

References

1. Holland JH (1992) Genetic algorithms. Sci Am 267(1):66–73
2. Deb K, Pratap A, Agarwal S, Meyarivan TAMT (2002) A fast and elitist multiobjective genetic algorithm: NSGA-II. IEEE Trans Evol Comput 6(2):182–197
3. Deb K, Goel T (2001) Controlled elitist non-dominated sorting genetic algorithms for better convergence. In: International conference on evolutionary multi-criterion optimization. Springer, Berlin, pp 67–81
4. Kukkonen S, Deb K (2006) Improved pruning of non-dominated solutions based on crowding distance for bi-objective optimization problems. In: 2006 IEEE International conference on evolutionary computation. IEEE, pp 1179–1186
5. Zitzler E, Brockhoff D, Thiele L (2007) The hypervolume indicator revisited: on the design of Pareto-compliant indicators via weighted integration. In: International conference on evolutionary multi-criterion optimization. Springer, Berlin, pp 862–876
6. Kursawe F (1990) A variant of evolution strategies for vector optimization. In: International conference on parallel problem solving from nature. Springer, Berlin, pp 193–197
7. Poloni C, Mosetti G, Contessi S (1996) Multi objective optimization by GAs: application to system and component design. In: Computational methods in applied sciences' 96, ECCOMAS'96. Wiley, 1–7

Chapter 5
Multi-objective Grey Wolf Optimizer

5.1 Introduction

Metaheuristics have become very popular in the last two decades. This class of problem solving techniques includes a wide range of algorithms to find reasonably good solutions for problems where deterministic methods are not efficient. Their name come from their mechanism, in which they do not required problem-specific heuristic information. Such methods are stochastic and consider problems as a black box.

Metaheuristics can be classified into two classes based on the number of solutions that they generate in each iteration. In the first class, one solution only is generated and improved until an end condition is met. In the second method, however, a group of solutions are used and improved for a given optimization problem. Algorithms in both class have their own advantages and drawbacks. The main benefit of algorithm in the first class is the cheap computational cost. This is because one solution is only evaluated using the objective function in each iteration. Another pros of such algorithms is the quick convergence rate. However, this leads the to a drawback, which is less exploration. One solutions might not be able to extensively explore the search space and is prone to be trapped in locally optimal solutions.

Algorithms using a group of solutions benefit from higher exploratory behaviour as opposed to the first class. This is because more solutions are able cover larger areas of search space. They can also share information about the shape and difficulty of the search space as well. As a drawback, however, each solution requires evaluation that should be done using the objective function. Therefore, the computational cost of such methods is a concern. Another drawback is related to space complexity. Algorithms in the second class required more memory to operator as compared to those in the first class.

Regardless of pros and cons of algorithms with single or multiple solutions, what makes them widely applicable is the gradient-free mechanism. As opposed the gradient-based algorithm, metaheuristics required little to no information about

S. Mirjalili and J. S. Dong, *Multi-Objective Optimization using Artificial Intelligence Techniques*, SpringerBriefs in Computational Intelligence, https://doi.org/10.1007/978-3-030-24835-2_5

the mathematical models of the problem. The only things that they need are the number of variables, the range of variables, the number of objectives, the number of constrains, and the objective function(s). They then constantly create random solutions for the problem and evaluate them. Of course, they are not complete due to the use of stochastic operators. However, they find reasonably good solutions for a given problem in a reasonable time.

Another classification of metaheuristic is based on the source of inspiration. The many recent metaheuristics can be divided into three classes: evolutionary, swarm-based, and physics-based. In the first class, the source of inspiration is evolutionary phenomena in nature. The preceding chapter introduced one of the most popular ones called Genetic Algorithm (GA). Other algorithms in this class are: Evolutionary Strategy, Differential Evolution, and Biogeography-based Optimization. Most evolutionary algorithms have four evolutionary operators: selection, reproduction, mutation, and elitism.

The second class includes algorithms that mimic the collective behaviour of creatures in nature that leads to intelligence and problem solving. One of the previous chapters covered Particle Swarm Optimization (PSO) as one of the most popular swarm intelligence technique. Other popular algorithms in this class are Ant Colony Optimization (ACO), Artificial Bee Colony (ABC) optimization, Whale Optimization Algorithm (WOA), and Dragonfly Algorithm (DA). The majority of swarm-based methods consider n-dimensional space and require solutions to move inside them when searching for the global optimum of an optimization problem.

This chapter first introduces Grey Wolf Optimizer (GWO) as one of the most recent swarm intelligence techniques proposed by the author of this book. The multi-objective version on this algorithm called Multi-Objective Grey Wolf Optimizer (MOGWO) is then discussed and tested in details since the scope of the book is on multi-objective optimization.

5.2 Grey Wolf Optimizer

The Grey Wolf Optimizer (GWO) was proposed in 2014 by the first author of this book [1]. This algorithm mimics the dominance hierarchy and hunting behavior of grey wolves in nature. Grey wolves live in one of the most organized natural groups called pack. Wolves in a pack are divided into four classes: alpha, beta, delta, and omega. The alpha is normally the strongest wolf that lead the pack in navigation and hunting. All wolves should follow alpha's order. In the next dominance lever, beta wolves help alpha in decision making and leadership. Omega wolves are the least powerful.

In hunt, all wolves follow the alpha order. Grey wolves tend to first chase the prey and circle around it. The team work gradually traps a prey. When chasing, encircling, and harassing, the prey gradually become tired. At this stage the final attack is done to kill the prey. This intelligence social behaviour allows grey wolves to forage preys bigger than themselves as well.

In the GWO algorithm, the power hierarchy of wolves in nature is mimicked by saving the three best solutions that this algorithm has found so far. Those solutions are equivalent to alpha, beta, and delta wolves. The rest of solutions are considered to be omegas. After defining the dominance level, the solution should be updated. In GWO, it is assumed that every grey wolf has a vector of position. There is not velocity vector and the solutions are updated my direct manipulation of the position vector. The proposed position updating equation for the solutions are as follows:

$$\overrightarrow{X}(t+1) = \overrightarrow{X_p}(t) - \overrightarrow{A} \cdot \overrightarrow{D} \tag{5.1}$$

where $\overrightarrow{X}(t+1)$ is the position of a grey wolf in $t+1$-th iteration, $\overrightarrow{X}(t)$ is position of the grey wolf at t-th iteration, $\overrightarrow{A}$ is a coefficient and $\overrightarrow{D}$ is the distance that depends on the location of the prey $(\overrightarrow{X_p})$ and is calculated as follows:

$$\overrightarrow{D} = \left| \overrightarrow{C} \cdot \overrightarrow{X_p}(t) - \overrightarrow{X}(t) \right| \tag{5.2}$$

$$\overrightarrow{X}(t+1) = \overrightarrow{X_p}(t) - \overrightarrow{A} \cdot \left| \overrightarrow{C} \cdot \overrightarrow{X_p}(t) - \overrightarrow{X}(t) \right| \tag{5.3}$$

$$\overrightarrow{A} = 2\overrightarrow{a} \cdot \overrightarrow{r}_1 - \overrightarrow{a} \tag{5.4}$$

$$\overrightarrow{C} = 2 \cdot \overrightarrow{r}_2 \tag{5.5}$$

where $\overrightarrow{a}$ is a parameter that balances exploration and exploitation. The random components of this equation are $\overrightarrow{r}_1$ and $\overrightarrow{r}_2$ that are randomly generated from the interval $[0, 1]$.

As discussed above the parameter a is the main mechanism to balance exploration and exploitation in GWO. In the original version of this algorithm, time-varying values are chosen to first explore (when $0 < a < 1$) and then exploit the search space (when $1 < a < 2$). The equation that require this parameter to be updated based on the current iteration is as follows:

$$a = 2 - t\left(\frac{2}{T}\right) \tag{5.6}$$

where t shows the current iteration and T is the maximum number of iterations.

The above equations can update the position of every gray wolf. They allow them to go 'around' other solutions in an n-dimensional search space just like how real grey wolves encircle a prey in the 3D space. To simulate how the position of each wolf is indicated using the alpha, beta, and delta wolves, the following equation was proposed in the original GWO:

$$\overrightarrow{X}(t+1) = \frac{X_1 + X_2 + X_3}{3} \tag{5.7}$$

where $\overrightarrow{X_1}$ and $\overrightarrow{X_2}$ and $\overrightarrow{X_3}$ are calculated with Eq. 5.8.

This equation shows that the new position of a wolf is the average of three components. These components are calculated as follows:

$$\vec{X}_1 = \vec{X}_\alpha(t) - \vec{A}_1 \cdot \vec{D}_\alpha$$
$$\vec{X}_2 = \vec{X}_\beta(t) - \vec{A}_2 \cdot \vec{D}_\beta \tag{5.8}$$
$$\vec{X}_3 = \vec{X}_\delta(t) - \vec{A}_3 \cdot \vec{D}_\delta$$

$\vec{D}_\alpha$, $\vec{D}_\beta$ and $\vec{D}_\delta$ are calculated using Eq. 5.9.

$$\vec{D}_\alpha = \left| \vec{C}_1 \cdot \vec{X}_\alpha - \vec{X} \right|$$
$$\vec{D}_\beta = \left| \vec{C}_2 \cdot \vec{X}_\beta - \vec{X} \right| \tag{5.9}$$
$$\vec{D}_\delta = \left| \vec{C}_3 \cdot \vec{X}_\delta - \vec{X} \right|$$

The GWO algorithm first starts the optimization process using a group of random solutions. This group is evaluate using an objective function. After knowing the quality of each solution, the best three are considered to be alpha, beta, and delta. The algorithm then iteratively updating the position of wolves while updating the time-varying parameters such as a. At any point in time, if a solution becomes better than alpha, beta, and delta, they have to be replaced by the new solution. The GWO algorithm stops after the satisfaction of the end criterion.

5.3 Multi-objective Grey Wolf Optimizer

The multi-objective version of GWO was proposed in 2016 by Mirjalili *et al.* to solve problems with multiple objectives [2]. Similarly to MOPSO [3, 4], MOGWO employs a archive to store the best non-dominated solutions throughout the optimization process. Storing non-dominated solutions in the archive should be done using the following rules:

- If the archive is empty and there is a grey wolf that is non-dominated, it should be added to the archive.
- If a solution in the archive is dominated with respect to a solution outside the archive (a grey wolf), it should be replaced with the new solutions immediately.
- If a solution is non-dominated in comparison with the solutions in the archive and we have enough space, the solution should be added to the archive.
- If a solution is non-dominated in comparison with the solutions in the archive and we have do not have enough space, on solution in the most crowded segments of the archive grid should be removed and the new solution should be inserted to the archive.

The archive mechanism is similar to that in MOPSO. It has a maximum size and should have two operators: archive maintenance and leader selection. In archive maintenance, solutions from crowded regions should be removed when the archive is full. The grid mechanism divides the objective space into segments. The crowdedness of each segment is defined by the number of solutions that it holds. Therefore, the probability of choosing the i-th segment to remove a solution from is calculated as follows:

$$p_i = \frac{n_i}{c} \tag{5.10}$$

where n_i indicates the number of non-dominated solution in the i-the segment and c is a constant that is normally set to 1.

This equation shows that the probability of choosing a crowded segment is high. If there is no non-dominated in the segment, the probability of removing a solution from it is equal to 0. The probability of selecting a leader from the archive is done in an opposite manner. The following equation show the equation used to find a suitable segment to choose a leader from:

$$p_i = \frac{c}{n_i + 1} \tag{5.11}$$

where n_i indicates the number of non-dominated solution in the i-the segment and c is a constant that is normally set to 1.

This equation shows that the fewer solutions in a segment, the higher probability of choosing the leader. In fact, a segment with no non-domineered is the most likely one that will be chosen by this mechanism, which is desirable since the aim here is to improve the coverage of solutions in the archive across all objectives. Note that n_i is added with 1 to prevent division by zero. Ac example of how the above two equations assign different probabilities to the segments can be seen in Fig. 5.1.

This figure shows that the probability values for segment increase as the number of solutions inside them decrease. The left figure shows that the probability of choosing the least crowded segment is equal to 1. Of course, there is no solution in this segment to choose as a leader. However, this can be fixed easily by considering only the segments with solutions inside them. An easier way is to increase the parameter c so that the least crowded region gives the probability less than 1. The key point here is that the method gives higher probability to the less crowded segments. Therefore, the MOGWO algorithm searches in those areas around the non-dominated solutions to find more non-dominated solutions and increase their overall distributions.

On the other hand, the right figure shows that the story is opposite when the archive is full. The probability of removing a solution from the archive is at the maximum level for the most crowded segment. This means that a solution will be chosen from this segment to accommodate a new one. In stochastic algorithm, we normally want to give a small probability of to crowded regions to. This will help exploration and avoiding locally optimal solutions. This can be done with considering something greater than $max(n_i)$ for the parameter c.

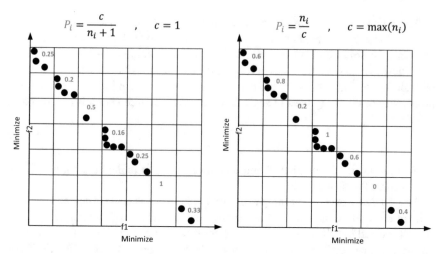

Fig. 5.1 (Left) The probability values for segment increase an the number of solutions inside them decrease. This is because we want to choose leaders from such regions to search around them and find more accurate solutions. This method also increases the coverage of solutions across all objectives over time. (Right) The probability of choosing removing a solution from segments increase proportional to the number of solutions insider them. This combined with the first method decreases the solutions from more dense regions and increases the density of the less dense regions

In the MOGWO algorithm, there are three leaders in each iteration: alpha, beta, and delta. In each iteration of the optimization, this algorithm uses the above leader selection technique to choose three non-dominated solutions. These are reference points to update all solutions in the population. After updating their position, there are inserted in the archive.

5.4   Literature Review of MGWO

The MOGWO algorithm has been widely used in both science and industry. Since the proposal, there has been several improvements and variants as well. This section provides a brief literature review of this algorithm.

5.4.1   Variants

This first version of MGWO uses an archive just like many other a posterior optimization algorithms. Once archive allows storing non-dominated solutions, and the algorithm keeps choosing leaders from this repository. However, there is a work that utilizes two archive called 2ArchMGWO [5]. In this method, one archive is used to improve exploration and the other one has been included to improve exploration. The authors considered different strategies to update the two archive and select leaders from them. Interested readers are referred to [5] for more details.

Another variant of MOGWO can be found in [6], in which the authors used a non-dominated sorting method proposed initially in the NSGA-II algorithm. In this method, all non-dominated solution in the archived are ranked based on the crowding distance which is another popular method to improve the coverage of solutions obtained by a posteriori optimization algorithm.

The last variant as March 2019, is the Multi-Objective Grey Wolf Optimizer based on Decomposition (MOGWO/D) [7]. This method is similar to Multi-Objective Evolutionary Algorithm based on Decomposition (MOEA/D), in which Pareto optimal solutions are approximated by defining a neighborhood among subproblems where the multi-objective problem is decomposed.

5.4.2   Applications

The MOGWO algorithm has been employed to solve a wide range of problems in both science and industry. Some of the popular and recent areas as of 2019 are as follows:

- Dynamic scheduling in a real-world welding industry [8]
- Scheduling problem in welding production [9]
- Multi-objective optimization of multi-item EOQ model with partial backordering and defective batches and stochastic constraints [10]
- Maximum power point tracking of wind energy conversion system [11]
- Wind speed multi-step forecasting [12]
- Designing photonic crystal filters [13]
- Designing photonic crystal sensors [14]
- Blocking flow shop scheduling problem [15]
- Integration of Biorefineries [16]
- Power Flow Problems [17]
- Wind power forecasting [18]
- Optimal Design of Large Mode Area Photonic Crystal Fibers [19]
- Ecological Scheduling for Small Hydropower Groups [20]
- Image segmentation [21]
- Enhancing participation of DFIG-based wind turbine in interconnected reconstructed power system [22]
- Extractive single document summarization [23]
- Virtual Machine Placement in Cloud Data Centers [24]

- Assignment and scheduling trucks in cross-docking system with energy consumption consideration and trucks queuing [25]
- Radiation pattern design of photonic crystal LED [26]
- Blocking flow shop scheduling problem [27]
- Task scheduling strategy in cloud computing [28]
- Multi-Objective Optimal Scheduling for Adrar Power System including Wind Power Generation [29]
- Estimation Localization in Wireless Sensor Network [30].

5.5 Results of MOGWO

This section provides several experiments to better understand the search patten and the impact of the main controlling parameters of MOGWO. Note that ZDT1 is used in all experiments [31].

5.5.1 The Impact of the Parameter a

In this experiment the impact of the parameter a on the performance of the MOGWO algorithm is investigated. Thirty wolves with 100 iterations are used to estimate the Pareto optimal solution set. Note that the parameter c is a random in [0, 2]. The result are provided in Fig. 5.2.

This figure shows that the MOGWO algorithm shows little convergence when a is equal to 0 or 0.1. This is because the movements around alpha, beta, and delta are very small for these values, which leads to very small movements of wolves in a search space. The MOGWO algorithm starts to show some converge when $a = 0.5$. However, all the solutions are gravitated towards the left corner of the front. A better convergence towards more Pareto optimal solutions can be seen when increasing the parameter a. When $a = 1$, the coverage of solutions are still not good. In the last two figures, however, the solutions are distributed uniformly across both objectives.

5.5.2 The Impact of the Parameter c

In this experiment the parameter a is set to be 0.5. This is because we want to see how much the parameter c can improve the exploration. The MOGWO algorithm is run six times while changing the parameter c. The results are presented in Fig. 5.3.

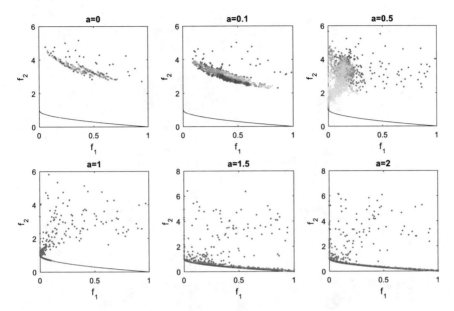

Fig. 5.2 he MOGWO algorithm shows little convergence when a is equal to 0 or 0.1. This is because the movements around alpha, beta, and delta are very small for these values, which leads to very small movements of wolves in a search space. The MOGWO algorithm starts to show some converge when $a = 0.5$. A better convergence towards more Pareto optimal solutions can be seen when increasing the parameter a. When $a = 1$, the coverage of solutions are still not good

This figure shows that the exploration is not broad when $c = 0$. The only random component that caused a little bit of exploration is the calculation of A. Figure 5.3 shows that the exploration of the search space increase when $c = 0.1$. However, they are many ares explored that might not be promising. In the rest of subplot, it is seen that the exploration is more directed. As the parameter c increases, the exploration becomes more directed. As always, too much exploration might result in degraded exploitation. There should be a good balance between exploration and exploitation.

Figure 5.3 shows that the a really smooth convergence and directed exploration can be seen when c is assigned with random values. This is why in the original version of the GWO and MOGWO algorithms, the parameter c is always a random number to provide stochastic behavior while provide micro switches between exploration and exploitation. In other works, the parameter a is linearly decrease to increase exploitation. The parameter c, however, cause exploration at different stages of optimization.

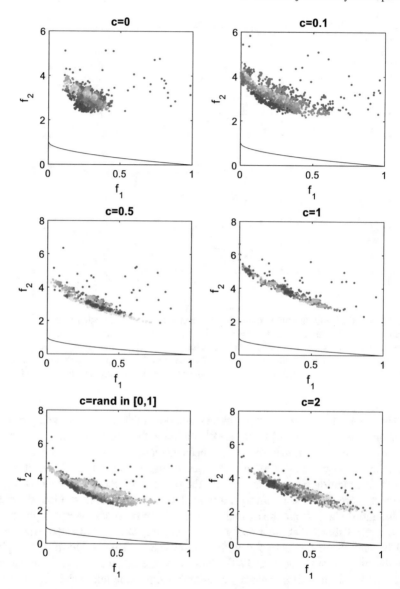

Fig. 5.3 The impact of the parameter c on the performance of MOGWO

5.6   Conclusion

This chapter introduced the MOGWO algorithm as one of the most recent a posteriori multi-objective optimization algorithm. The GWO algorithm was first presented since MOGWO uses most of the GWO's search mechanism. After than, the modification to GWO that led to MOGWO was presented. It was discussed that MOGWO has an archive, an archive controller, and a leader selection mechanisms. The chapter also included testing the performance of the MOGWO algorithm on a test function while changing the main controlling parameters: a and c.

References

1. Mirjalili S, Mirjalili SM, Lewis A (2014) Grey wolf optimizer. Adv Eng Softw 69:46–61
2. Mirjalili S, Saremi S, Mirjalili SM, Coelho LDS (2016) Multi-objective grey wolf optimizer: a novel algorithm for multi-criterion optimization. Expert Syst Appl 47:106–119
3. Coello CC, Lechuga MS (2002) MOPSO: a proposal for multiple objective particle swarm optimization. In: Proceedings of the 2002 congress on evolutionary computation, CEC'02 (Cat. No. 02TH8600), vol 2. IEEE, pp 1051–1056
4. Mostaghim S, Teich J (2003). Strategies for finding good local guides in multi-objective particle swarm optimization (MOPSO). In: Proceedings of the 2003 IEEE swarm intelligence symposium, SIS'03 (Cat. No. 03EX706). IEEE, pp 26–33
5. Nuaekaew K, Artrit P, Pholdee N, Bureerat S (2017) Optimal reactive power dispatch problem using a two-archive multi-objective grey wolf optimizer. Expert Syst Appl 87:79–89
6. Jangir P, Jangir N (2018) A new non-dominated sorting grey wolf optimizer (NS-GWO) algorithm: development and application to solve engineering designs and economic constrained emission dispatch problem with integration of wind power. Eng Appl Artif Intell 72:449–467
7. Zapotecas-Martnez S, Garca-Njera A, Lpez-Jaimes A (2019) Multi-objective grey wolf optimizer based on decomposition. Expert Syst Appl 120:357–371
8. Lu C, Gao L, Li X, Xiao S (2017) A hybrid multi-objective grey wolf optimizer for dynamic scheduling in a real-world welding industry. Eng Appl Artif Intell 57:61–79
9. Lu C, Xiao S, Li X, Gao L (2016) An effective multi-objective discrete grey wolf optimizer for a real-world scheduling problem in welding production. Adv Eng Softw 99:161–176
10. Khalilpourazari S, Pasandideh SHR (2018) Multi-objective optimization of multi-item EOQ model with partial backordering and defective batches and stochastic constraints using MOWCA and MOGWO. Oper Res 1–33
11. Kahla S, Soufi Y, Sedraoui M, Bechouat M (2017) Maximum power point tracking of wind energy conversion system using multi-objective grey wolf optimization of fuzzy-sliding mode controller. Int J Renew Energy Res (IJRER) 7(2):926–936
12. Liu H, Duan Z, Li Y, Lu H (2018) A novel ensemble model of different mother wavelets for wind speed multi-step forecasting. Appl Energy 228:1783–1800
13. Mirjalili SM, Merikhi B, Mirjalili SZ, Zoghi M, Mirjalili S (2017) Multi-objective versus single-objective optimization frameworks for designing photonic crystal filters. Appl Opt 56(34):9444–9451
14. Safdari MJ, Mirjalili SM, Bianucci P, Zhang X (2018) Multi-objective optimization framework for designing photonic crystal sensors. Appl Opt 57(8):1950–1957
15. Yang Z, Liu C (2018) A hybrid multi-objective gray wolf optimization algorithm for a fuzzy blocking flow shop scheduling problem. Adv Mech Eng 10(3):1687814018765535
16. Punnathanam V, Sivadurgaprasad C, Kotecha P (2016) Multi-objective optimal integration of biorefineries using NSGA-II and MOGWO. In: 2016 International conference on electrical, electronics, and optimization techniques (ICEEOT). IEEE, pp 3970–3975
17. Dilip L, Bhesdadiya R, Trivedi I, Jangir P (2018) Optimal power flow problem solution using multi-objective grey wolf optimizer algorithm. In: Intelligent communication and computational technologies. Springer, Singapore, pp 191–201
18. Hao Y, Tian C (2019) A novel two-stage forecasting model based on error factor and ensemble method for multi-step wind power forecasting. Appl Energy 238:368–383

19. Rashidi K, Mirjalili SM, Taleb H, Fathi D (2018) Optimal design of large mode area photonic crystal fibers using a multiobjective gray wolf optimization technique. J Lightwave Technol 36(23):5626–5632
20. Wang Y, Wang W, Ren Q, Zhao Y (2018) Ecological scheduling for small hydropower groups based on grey wolf algorithm with simulated annealing. In: International conference on cooperative design, visualization and engineering. Springer, Cham pp 326–334
21. Oliva D, Elaziz MA, Hinojosa S (2019) Image segmentation as a multiobjective optimization problem. In: Metaheuristic Algorithms for image segmentation: theory and applications. Springer, Cham, pp 157–179
22. Falehi AD An innovative OANFIPFC based on MOGWO to enhance participation of DFIG-based wind turbine in interconnected reconstructed power system. Soft Comput 1–17
23. Saini N, Saha S, Jangra A, Bhattacharyya P (2019) Extractive single document summarization using multi-objective optimization: exploring self-organized differential evolution, grey wolf optimizer and water cycle algorithm. Knowl-Based Syst 164:45–67
24. Fatima A, Javaid N, Anjum Butt A, Sultana T, Hussain W, Bilal M, Hashimi M, Ilahi M (2019) An enhanced multi-objective gray wolf optimization for virtual machine placement in cloud data centers. Electronics 8(2):218
25. Vahdani B (2019) Assignment and scheduling trucks in cross-docking system with energy consumption consideration and trucks queuing. J Cleaner Prod 213:21–41
26. Merikhi B, Mirjalili SM, Zoghi M, Mirjalili SZ, Mirjalili S (2019) Radiation pattern design of photonic crystal LED optimized by using multi-objective grey wolf optimizer. Photonic Netw Commun 1–10
27. Yang Z, Liu C, Qian W (2017) An improved multi-objective grey wolf optimization algorithm for fuzzy blocking flow shop scheduling problem. In: 2017 IEEE 2nd advanced information technology, electronic and automation control conference (IAEAC). IEEE, pp 661–667
28. Sreenu K, Malempati S (2018) FGMTS: fractional grey wolf optimizer for multi-objective task scheduling strategy in cloud computing. J Intell Fuzzy Syst (Preprint) 1–14
29. Mohammedi RD, Mosbah M, Kouzou A (2018) Multi-objective optimal scheduling for adrar power system including wind power generation. Electrotehnica, Electronica, Automatica 66(4):102
30. Thom HTH, Dao TK (2016) Estimation localization in wireless sensor network based on multi-objective Grey Wolf optimizer. In: International conference on advances in information and communication technology. Springer, Cham, pp 228–237
31. Zitzler E, Brockhoff D, Thiele L (2007) The hypervolume indicator revisited: On the design of Pareto-compliant indicators via weighted integration. In: International conference on evolutionary multi-criterion optimization. Springer, Berlin, pp 862–876

Printed in the United States
By Bookmasters